To the memory of
Maria Polly Bingham Ball,
mother of five sons,
who imbued them with the idea
of working together,
an idea which led to
Ball Corporation.

Left to right: William C. Ball, Frank C. Ball,
Lucius L. Ball (seated), Edmund B. Ball, and George A. Ball.

BALL
CORPORATION

The
First
Century

by Frederic A. Birmingham

CONTENTS

ILLUSTRATIONS

ACKNOWLEDGMENTS

The role of the historian is at best seen through a glass darkly.

It is put in rather dour perspective by none other than Dr. Nicholas Murray Butler, the famous president of Columbia University in the 1920s. When an associate alleged that the first use of an anecdote constitutes originality, the second plagiarism, the third a lack of originality, but the fourth "a drawing from common stock," the great scholar added briskly:

"Yes, and in the case of the fifth it is *research*."

Smarting somewhat at this after more than a year spent in researching this centennial volume, this author was somewhat reassured by the more kindly opinion of President G. Wayne Glick of Keuka College, a school which plays a part in our book:

"The 'ideal' historian can be described as an artist; he is one who possesses a broad and generous human viewpoint, a catholicity of spirit, and at the same time the specific skills of the historian's workshop: attention to detail, a critical eye, the patience and hard work required for the discovery and analysis of the sources, and the power of discernment."

Hovering somewhere between these two ends of the same spectrum, and hoping to combine them to produce a shaft or two of light, there is also the author's viewpoint to add to the perspective.

To be successful the historian must have a story worth telling to begin with. And the historian of a corporation's life is doubly challenged in this respect. On the shelves of every library there are a few absorbing chronicles which find a fitting place in the larger history of our nation and of our times. Others are written in a style so persuasive and at such a distance from actual fact that they should be properly catalogued under "fiction." Fortunate is the corporate historian who has the drama of great lives at his disposal to blend in with the romance of pure fact, and make the most of Emerson's opinion that "there is properly no history, only biography."

So, fortunate I.

The Ball Corporation story is one of colorful human enterprise amply described in its own records, and frequently in the public press, but never really brought together before in the entirety of its singular human elements. Add to this, achievements which have brought a tiny business originated on a few borrowed dollars by five young brothers to a place among the top 500 U.S. industrial corporations today and we have a tale which demands to be told. How could anyone resist the story of a corporation which has contrived to fling Space Age hardware into the heavens as successfully as to produce a simple Mason jar immaculately so as to become a thing of beauty?

In truth, this author inherited a story of human drama and technological advancement spread out over years which are the most prodigious in our nation's history. With such inner momentum the story almost tells itself. Here are moments of love and cheer, and tragedy faced with quiet dignity. Courage tested in crisis. Crushing failure, soaring success. The fulfillment of one family's belief in humanity and divine guidance. It has been my privilege to interview members of the Ball family in all parts of the country and to hear from their own lips reminiscences in which "founders" suddenly turn into fathers and uncles, and newspaper headlines of Ball entrepreneurial stories suddenly are reduced to personal chitchat over the family dinner table. I was enabled to delve into the past with veteran members of the company, with a "no holds barred" rule in effect as to what they might elect to say. Doors were opened to the most confidential records of the corporation. And beyond that into the private archives of the Ball family.

In that solitary exploration among the Ball memorabilia I must confess to sudden encounters of a very moving nature. A son of one of the founders, serving overseas in World War I, sends his mother a steady stream of photo-postcards, proudly depicting his rising rank, yet managing always to include manly words of devotion which must have made her heart very glad. Photographs of tiny children—curly haired, dimpled and wimpled and innocent as angels, well almost—whom I know now as adults. Dance programs from days long past when each radiant girl carried a tiny book in which to inscribe the names of dancing partners from every dreamy waltz and heady polka. Wedding invitations. Herbert Hoover writes a personal letter to George A. Ball from the White House. Another note is from Oyster Bay, New York, as Theodore Roosevelt has a few thoughts for G.A.

Our history suddenly becomes the biographies of all those who have supplied the lifeblood and the pulsebeat for the curious phenomenon known as a corporation. In the case of the Ball Corporation, one of its fascinating traits across the span of the century we record is the unbroken continuity by which the enduring philosophy of its 19th century founders guides their descendants through the undreamed-of complexities of 20th century expansion. It is an exciting story of men who were no less pioneers than those on the wagon trails trekking west. They pioneered a new kind of society by creating technologies to provide a better way of life for the average citizen. It is they, and the hundreds of thousands of others whose lives they influenced, who are the real authors of this story.

I am only grateful for the opportunity to weave together the illustrious course of those lives and those years, a quiet tale of astounding reality.

So many individuals contributed to this volume that it would be impossible to include the names of all who helped make it come to pass. It was a group effort, and our thanks are due to all who made it possible. But there are some who cannot go unmentioned because of special contributions without which this book could not be.

The guiding light of the project was Edmund F. Ball, who shepherded it along gently from the first word to the last and whose own passages are gems of reminiscence, bringing alive for us in loving detail those members of his family who have passed on. Elsewhere in our research, Ed's encyclopedic memory and corporate insights lent vision and clarity to these pages. He was ever open minded and objective, sometimes accepting data which might reveal human flaws—as we now perceive them in hindsight—or spotlight imperfect company procedures. His day-to-day interest over the extended time it took to create this history was an inspiration and guide to us all.

John W. Fisher, Richard M. Ringoen and Alexander M. Bracken were others who contributed invaluable insights, particularly on company operations in recent years and concerning future directions. The daughters of the five founders of loving memory lent a special charm and grace to our story. Our warm thanks are due to Miss Elisabeth Ball, Mrs. Rosemary Ball Bracken, Mrs. Janice Ball Fisher, Mrs. Lucy Ball Owsley and Mrs. Margaret Ball Petty. Also Mr. and Mrs. William H. Ball, who gave us glints of family dignity and sprightliness as they recalled the past.

Mr. John P. Collett of Indianapolis, financial adviser to the company over long and eventful years, provided memorable images of corporate survival and growth.

Vern C. Schranz, secretary of the corporation, was a tower of strength and patience, guiding the work over innumerable sessions during its preparation, and offering counsel worthy of a Tallyrand and Solomon combined. John J Pruis, vice president of corporate relations, gave us a scholarly and articulate presence which future historians may well note. William F. Brantley was ever a helpful guide and critic through much of the effort, finding a path through the maze of factual entanglement whenever the way became something less than distinct.

A blueprint for the history was contained in the voluminous and detailed notebooks which the late George E. Myers, former treasurer and board member of the corporation, over four years painstakingly gathered in the hope that someday his data would provide for a more literary history of the company. About Myers' work Ed Ball once said: "It is important that this factual history of the company be preserved and brought up to date." We bow in gratitude for the groundwork broken by the indefatigable Mr. Myers.

In the course of researching our history, we visited many Ball Corporation facilities, and interviewed executives and line workers literally from coast to coast. Once more we are compelled to offer them our gratitude as parts of the Ball team, rather than attempt to list them all. However, certain individuals—some of whom are retired, and others still on the firing line—made notable contributions which deserve our special thanks. The archives of the corporation have retained recordings of our conversations with them, and they will be duly preserved. Their esteemed names: Otto Edwin Bartoe, Joe B. Brown, Maurice C. Clark, Richard T. Ekrem, Rodney Ford, R. Arthur Gaiser, P.R. (Phil) Goetz, Clarence Hamilton, Burnham B. Holmes, Ruel C. Mercure, Jr., Robert H. Mohlman, R.H. Morehouse, J.N. Ruthenburg, George A. Sissel, John P. Stevenson, Arthur M. Weimer.

Others who helped us generously along the way with suggestions and personal reminiscence were Jack Ferris, editor emeritus of the *Muncie Star*, and Dick Greene, *Muncie Star* columnist, whose off-the-record viewpoints were invaluable in sorting out the human elements in our family history.

As companions in research, we hail with respect Bill Spurgeon and other staffers of the *Muncie Star, Evening Press,* and *The Indianapolis News.* We are equally in debt to the staff of the Muncie Public Library for helpfulness and courtesy beyond the call of duty. Special thanks for reprint rights are due to Dodd, Mead & Co. (*How Dear To My Heart* by Emily Kimbrough); *Na-*

tion's Business Magazine; Scripps-Howard Newspapers; *BusinessWeek* Magazine (McGraw-Hill, Inc.); and Harcourt Brace Jovanovich (*Middletown* and *Middletown in Transition* by Robert S. and Helen Lynd).

As the book progressed, the author became deeply indebted to many at *The Saturday Evening Post* for their valuable and unceasing support. To Beurt R. SerVaas, chairman of Curtis Publishing, for countless Sunday afternoons, with other business pressures and a chance for relaxation insistently at his elbow, as he diligently pored over the manuscript and added his own knowing and refining touches. To Dr. Cory SerVaas, editor and publisher of *The Saturday Evening Post*, for her inspiration and encouragement throughout the long process of bringing together the disparate elements of our history. To Michael Hayes for his splendid design of this book and its contents. To Tom O'Neil, my brilliant special assistant on the project, for his unflagging research and championing of *le mot juste.* To Steve Frye, whose special province was the art direction of the illustrated pages in the book, almost a book within-a-book. And to our proofreaders and typists, who never wavered as they worked their way through myriad versions of the manuscript, amounting to quite a few volumes in terms of total words, doing this as a work of love above their regular duties on the *Post* itself. Hail to them all.

—FREDERIC A. BIRMINGHAM

BALL CORPORATION

*The
First
Century*

"It may be self-serving to say we know what we are doing, but we really do. We plan. We seek out sound, growing markets. We organize. We anticipate. We budget carefully. We work hard. We are productive. We manage change rather than let change manage us. If that is special or unusual, then perhaps we have special or unusual management skills."

—John W. Fisher
*Chairman of the board
and chief executive officer
Ball Corporation*

INTRODUCTION
by Edmund F. Ball

My recollections of over 50 years' association with the company lie essentially in two categories—people and events.

Contrary to a popular misconception that corporations are cold, impersonal institutions possessing neither heart nor soul, each is an ever-changing mosaic composed of the composite characteristics of hundreds, maybe thousands, of individuals who have participated in the shaping of its image.

This book begins at the bottom line, the corporation in 1980 at the end of its first 100 years. It tells the story of how the company got there through trials and tribulations, the thrills of success, the agonies of failure, opportunities capitalized on, opportunities lost. And reflecting the historic policy of the company from its beginning—to plan ahead—the book dares to look into the future.

Mostly it's about people, the five brothers who minded their mother's admonition to always stick together in whatever they did. It tells about those with whom the brothers were associated in their earliest days. It speaks of those who followed, building on the foundations so firmly and so successfully laid, who have continued as well the principles, philosophies, and policies the brothers established by writ and by example.

It is a fascinating exercise, even though it may seem somewhat futile, to speculate on what might have been the results had opposite courses been followed, or if happenstances had developed differently at certain critical periods in the company's history. Carefully laid plans sometimes succeeded or sometimes failed. Major decisions may have worked out well—or badly.

The history of the world has turned on events that at times might have seemed unimportant. Luck, fate, circumstances have played their parts. Whims, personalities, temperaments, prejudices of individuals have all helped to weave the intricate pattern that becomes the tapestry of history.

The same factors shape the destiny of a business enterprise. Far from being a faceless, soulless institution, a corporation is a living thing, the result of many events, the product of so many who have played a part in shaping its destiny. To mention a few does injustice to many, but there is no practical alternative. Let's begin by remembering some of them.

Of course, at the beginning, there were the five able, energetic brothers who founded the company. Their talents and personalities were remarkably well balanced. Frank C. was a born leader, a strong, dynamic, shrewd businessman. He guided the company's destiny literally to his deathbed in 1943 at the age of 85, concluding a remarkable business career of 63 years as founder and president. Edmund B. Ball, my father, was a great humanitarian. He loved people and liked to work with his hands, planning and building—a balance wheel for the enthusiasm and energies of his brothers. His forward-thinking plans led to some of the brothers' finest achievements even after his untimely death in 1925. William C. was an exceptional salesman and traveler, elocutionist and raconteur. The oldest of the brothers, Lucius L., fulfilled his lifelong ambition (after he had seen to it that his younger brothers and sisters were educated and established) to study and become a physician at the age of 40. The youngest brother, George A., lived to the ripe old age of 92, keen, active, alert to the very end in 1955. During his long and active lifetime, he served the company as bookkeeper, secretary, treasurer, vice president, president, and board chairman. He participated in all the changes from kerosene cans and fruit jars to the threshold of the space age.

There had been the father, "Squire" Lucius Styles Ball, farmer, inventor, respected citizen, dearly beloved by his family and his devoted wife, Maria. He had instilled confidence in his sons, and assured them that they would succeed in whatever in life they undertook. The mother had given her children love, inspiration, and direction. The two sisters, Lucina and Frances, later distinguished themselves in the field of education.

Uncle George H. Ball, Baptist minister, educator, and founder of educational institutions, was the young men's mentor, business adviser, and confidant.

Moving to Muncie when the Buffalo Glass Plant burned was the able Peter Menard, first plant superintendent; and with him glassblowers like Adam Traub, Adolph and Henry Miller, "Cap" Corey, Billy Bryan, Mike Shepner, Bill Gets, William Thompson the blacksmith, and gentlemanly Charles Dolloway. There was Ham Dunnington who could fix anything from a watch to a steam engine; and the mechanical and engineering gen-

ius, Cousin Al Bingham, who designed and built machinery far ahead of the times. Fred E. Jewett, protege of E. B. Ball, for 55 years a loyal employee, and for a quarter century or more, glasshouse superintendent. He was my mentor in the early days. M. L. Hageman, austere disciplinarian, was bookkeeper, auditor, office manager, teacher of office etiquette and business ethics. And there were two traveling salesmen, Mel Doran and "Colonel" Charlie Lincoln, true "ambassadors of commerce." And Charlie's sister—"cousin" or "aunt" (whichever might be appropriate)—Mary, cashier, scrutinizer of salesmen's expense accounts. In manufacturing were the Scott brothers, Jimmy and Dave, assistants to Fred Jewett; the Ludington brothers, Lawrence and Will; the Wallace brothers, Bert, Will, and Ray; the Dungans, Ernest and Arthur; and so many more. Dan Hinchcliff brought with him from La Harpe, Kansas, the secrets of rolling strip zinc. Such men as Tom Smock, Spanish-American War veteran, at Wichita Falls, Texas; the fiery Scotty Davidson at Hillsboro, Illinois; and the irascible, cantankerous C. H. Inglish at Okmulgee, Oklahoma, managed the outside plants with iron hands. Lovely Gertrude Barrett served the company and family as switchboard operator, receptionist, confidante, and private secretary for a lifetime. There was Frank E. Burt, office boy to sales manager in 60 years of service, who played a major part in determining policies and developing plans which made Ball fruit jar products and home canning supplies known throughout the world. George E. Myers came to the company as accountant and auditor with his associate Roy Hill, and accounting systems began to be formulated. He was the first person not a member of the family to become an officer and director. It was he who, after retirement, collected papers and documents, and put them together with notations and personal recollections as permanent records of the company, supplying a basis of background material for this book. Earl Milner, engineer, designed plants and systems, many still in existence; Walter Sterrett, molds and machinery. Charlie Austin ran the paper mills; and Bill Exton, beginning by manufacturing rubber fruit jar sealing rings with second-hand machinery, started a rubber operation that for 40 years was an important division of the company. G. Fred Rieman left Anchor Hocking and joined the company in 1939. He was the first person outside the family to be named a vice president when he was put in charge of glass container operations. Hugh Crawford in the late 1940s built and managed the California glass plant for many years.

And what about the family at the half-century mark? "W. C.," the first of the brothers to break the circle, died in 1921; "E. B." in 1925; and the doctor in 1932. E. Arthur, oldest son of "F. C.," and William H., son of "W. C.," joined the company at the conclusion of World War I. Arthur—affable, gregarious, a bit of a nonconformist—assigned to sales, was never too happy in the business. Nevertheless, he returned to it after serving again in World War II and died unexpectedly in 1947 at Millville, New Jersey, while serving as manager of the Friedrich and Dimmock Company, manufacturing specialized glass fibers. Will —energetic, unpredictable, impulsive—directed his enthusiasms toward the development of the zinc and paper operations with a considerable degree of success. Frank E., youngest son of "F. C.," talented and personable, who would undoubtedly have gone far in business and the company, as well as in civic leadership, was lost tragic-

ally in an airplane accident in 1936. The diligent, conscientious, hard-working Fred J. Petty, son-in-law of Frank C., was corporate secretary and manager of the Commercial Glass Sales Department. He could well be considered the founder of what has become such a major division of the company. He died unexpectedly while at work in the office in 1949. "F. C." 's third son-in-law, Alec Bracken, entered the organization as legal counsel and graduated into labor relations as the economy moved out of the depths of the Great Depression and labor unions began to flex their muscles. Later, of course, he served as general counsel and in many other capacities, eventually becoming chairman of the board. During the critical years of World War II, the first of "F. C." 's sons-in-law, Alvin M. Owsley—World War I colonel, a founder and early commander of the American Legion, soldier, politician, lawyer, diplomat —moved from Texas to Muncie to assist in coping with the company's massive problems.

I joined the organization in 1928 as a laborer in the glass factory and moved through various positions, never dreaming that one day I would succeed to the presidency of the enterprise. World War II called me to military service in 1941 and Arthur soon followed. John W. Fisher, who would eventually become chairman of the board and chief executive officer, joined the organization in that same year, and I had about two weeks to teach him everything I knew about the glass business (which didn't take very long and undoubtedly could have been accomplished in a much shorter length of time). The trying years of the war passed, leaving an organization decimated by attrition.

Russell G. Isenbarger, president of the then Merchants Trust Company of Muncie, which held major holdings of Ball Brothers Company stock in its trust accounts, became the first "outside" director of the company. Following shortly after Isenbarger came John P. Collett, my lifelong friend from college days, the company's first "outside" director to act as financial adviser and counselor. It was he who negotiated the company's first "modern" major financing at a crucial time in its history.

Later, during the reconstruction period, appeared the controversial, abrasive Duncan C. Menzies, who, in spite of all his shortcomings and exasperating assumptions and hypotheses, brought expertise and a certain degree of sophistication, not previously available, to the organization. There was W. C. Schade whose relatively short tenure as chief executive officer brought definitiveness to organization, approaches to acquisitions, research and development programs, and systems and procedures of great value to a fast-growing enterprise. From an era of change through crisis and necessity, the company was emerging into one of orderly planning. More euphoniously, I've referred to it as a "period of conservative opportunism."

I've mentioned only a few of the hundreds upon hundreds whose lives and services have shaped the destiny, the character, the image of what is called by the grossly inadequate name "a corporation."

In this brief review, we must also consider some of the principal events and major decisions that brought the company to its present position 100 years after its humble beginnings in Buffalo, New York. What might have resulted, for example, had not two fires in those early days changed the entire direction of the brothers' efforts? What if their Uncle George had not encouraged Frank and

Ed to continue their search for business opportunities in the face of adversities? What if Frank C. Ball had not become bored while looking over Bowling Green, Ohio, as a possible plant location and had not, at the precise psychological moment, received that provocative telegram from a little town in Indiana called Muncie, of which he had never heard? What if the Ball brothers had retained their interest in the Muncie-based Inter-State Automobile Company when Durant acquired it to make a major part of what is now General Motors? Or what about the acquisition of the La Harpe, Kansas, Zinc Mill in 1913? And the "Toledo Decree" that enjoined the company and its officers *forever* from pursuing certain courses within the glass container industry, thereby completely altering its anticipated growth pattern? What about the acquisition of Control Cells in Boulder, Colorado, in 1956, followed in 1962 by the traumatic decision to close the Muncie Glass Plant? Or, at an earlier date and even more significantly, what if the shareholders and directors at that crucial meeting in 1947 to decide the future course of the company had selected any of the other alternatives? Had any of the other alternatives considered been chosen, there would be no book, for there would have been no company remaining about which to write.

While I was still in college and shortly after I joined the company in 1928, there were many long and serious meetings to consider merger opportunities with Hazel-Atlas Glass Company, Owens-Illinois Glass Company, and others. Each opportunity was debated at great length. Each time the decision rested upon what seemed to be the best in the final analysis for the family and the longtime employees of the company. It was a fierce spirit of pride in having created a successful enterprise that would preserve and perpetuate the family name that should not be lost through sellout or merger that became the final determining factor.

Frank C. Ball, company president and founder, always made the final decision after considering all the pros and cons. He was a strong, forceful leader, but in reaching his decisions he thought first of family ties and the best interests of friends and associates. It was that philosophy that preserved the company's autonomy, and has kept members of the Ball family in a position of principal ownership, with third-generation members now active in the organization.

Fred Birmingham's challenge has been to document the history and portray the character of this great company at its century mark. He does so by reflecting on the sequence of events as they occurred and by acquainting his readers with those who individually and collectively participated in them and caused them to happen.

It is an important documentary of Americana, well worth contemplating and preserving for posterity.

Chapter 1

MUNCIE, 1980: THE VIEW FROM HERE

Tucked away in the subtitle of this book, "The First Century," is a revealing little word of prophecy.

That word is "First."

Its message: "The best is yet to come."

On that note of expectancy and determination, we perceived that the best way to tell this Ball story was from the present to the past because at this moment in time, we can take stock of where the company stands today, and better understand the inner meanings of what really took place over those hundred years of dedicated building.

As Ed Ball remarked, a corporation is not a soulless institution. Ball Corporation is a striking example. Here in the international headquarters of the company, fronting on High Street in Muncie, Indiana, there is ever-present evidence that "people count."

This corporate office building, which has won awards and citations for excellence in design since its opening in 1976, leaves nothing to be desired in terms of both efficiency and aesthetics, quite in keeping with a company which, in the years from 1970 to 1980, increased its net sales from $170.9 million to beyond the half-billion dollar mark, with great expectations of further gains in employment and sales in years to come.

But something of the character of this unusual corporation may also be gleaned from three seemingly incidental facets of this handsome new home office. At one spot on the floor of the lobby, a visitor comes upon a small cluster of tiles of antique finish, set in the ultra-modern composition of the new flooring. The tiles were saved from the old offices, located next to the former glass plant on Macedonia Avenue. Elsewhere in the lobby, close to the bank of elevators, the visitor may find some "old" bricks set in the modern wall design. They are survivors from the Muncie Central High School building which stood for decades at this address before being demolished to make way for the Ball building.

And, as a last touch to this cherished sense of continuity revealed everywhere in this company, we find in the archives a photograph of an antique automobile, with Ed Ball and John Fisher enthusiastically aboard, making the symbolic journey from the old Macedonia offices to the new address. Even the antique automobile was historically significant. It was a 1915 Inter-State, originally made in Muncie by a company organized by local residents, among them the esteemed Ball brothers. An interesting footnote to that association is the fact that after local interests in the company were sold, the spare parts were moved to a vacant building on the Ball Brothers' property and sold from there. A man named William "Cliff" Durant then took the Inter-State to Detroit where it became an important segment of what eventually became General Motors.

What we discover, then, in these executive offices is an intimate connection with the past, which is rare indeed in large corporations, however vigorous and self-renewing they may be. Looking out the window, we see a sign bearing the name, appropriately, of A. E. Boyce, the son of the man who, in 1886, invited the Ball brothers to settle in Muncie with their glass business, offering them enticements of $5,000, free land and natural gas. Many of the people in Muncie today belong to families which for generations have worked for Ball enterprises, and "family" is not too intimate a word to use in connecting this community and this corporation.

We are discussing the building of the new headquarters in Muncie with John W. Fisher, chairman and chief executive officer of Ball Corporation, and Richard M. Ringoen, its president since 1978.

Dick notes that "there were a lot of people who were quietly urging the executives to move to another location, possibly to a larger city like Chicago. But Ball management did not buy the idea. They foresaw what is now commonplace, with people moving out of Chicago and New York City into small communities not unlike Muncie. Said one executive: 'We have found we have been able to attract good people here. Muncie is much more attractive now in the minds of the doubters than it was back in '72 or '73 when a firm decision to remain here was made.' "

"It was simple," says Fisher, shaking off Dick Ringoen's suggestion that John Fisher was probably singlehandedly responsible for keeping the headquarters here. "Our homes were here. We

have good schools. We had no serious problems of violence or other problems which are prevalent in some of the metropolitan areas. Most of the key people in the company enjoy and appreciate living in a small-town atmosphere. Ball State University is here, with its cultural and educational advantages. We also felt that our new office would accommodate our anticipated space requirements for many years, as well as be an attractive building and an asset to downtown Muncie."

Recently, at the groundbreaking for the new Colorado Office Center in Westminster, Colorado, John remarked to those attending the event that when Ball first came to Colorado in 1956, "We made an investment in people and ideas, not in brick and mortar." He admitted, though, that the growth of Ball's Colorado operations from three employees to nearly 2,000 had since necessitated considerable brick and mortar, and believed that the Colorado Office Center would solidify Ball's presence a thousand miles from Muncie, in the shadow of the Rockies where once William C. Ball dreamt of building a glass plant.

"People and Ideas."

That is the tipoff on Ball's corporate philosophy. John was to state his own interpretation when asked why the company seemed to be thriving again in the 1970-1980 decade:

"It may be self-serving to say we know what we are doing, but we really do. We plan. We seek out sound, growing markets. We organize. We anticipate. We budget carefully. We work hard. We are productive. We manage change, rather than let change manage us. If that is special or unusual, then perhaps we have special or unusual management skills.

"The strength of our organization is and has been for years our ability to sell industrial products. Once we have secured an order with a volume opportunity, we've been able to perfect the engineering of the product, the production methods, and to price our efforts very competitively. We seem to have a propensity to take on an opportunity and then know what to do with it—even divestiture if we find it does not meet our expectations. We make and sell large quantities of an item. That's what we seem to do best."

John was only the company's fifth president since Frank C. Ball was its first leader in those early Buffalo days of 1880. It makes a remarkable story, which we shall follow in chapters to come, as these people of ideas and competence each forged a personal link in the human chain which provided for this company's rare longevity. Each had to provide leadership in the face of changes in technology and science, national growth, and social change in a world which was literally transformed in the century to follow.

John's special world at Ball in 1970, when he became president and chief executive officer, was one of diversity bewildering to the outside viewer. Time and time again in this story of the Ball endeavors we shall discover that what seems at first glance to be a completely unrelated venture on the part of the corporation actually had its roots in a preceding enterprise. Even in 1980, when this company would be categorized by its top executive as a packaging company with a high-technology base, the mind flicks back to Lucius Styles Ball, father of the five founding brothers, who in 1876 took out a patent on a new kind of package for protecting eggs during shipment. Maybe things haven't changed that much in Ball Corporation after all.

But F. C. would certainly have thought so, had he been there in 1970 to have experienced what John Fisher was to preside over. Ball plants in that year were spread out all over the world. Ball glass plants were located from coast to coast. Muncie still had the office headquarters and a plant for manufacturing home canning fittings, along with research, development and engineering facilities.

Out in suburban Denver, Jeffco Manufacturing Company, a subsidiary, was making tin plate and aluminum beer and soft-drink cans and bottle crowns. In Greeneville, Tennessee, Ball had recently built the world's largest and most modern facility for continuous casting, rolling, and fabricating zinc, replacing the rolled zinc and fabricating operations previously conducted in Brooklyn, New York, Muncie and Greencastle, Indiana.

Caspers Tin Plate Company, a Ball subsidiary, was producing lithographed and coated metals in Chicago, along with its division, Lafayette Steel and Aluminum, a metal service center. Ball was then making rubber products at Chardon Rubber Company in Chardon, Ohio, and Industrial Rubber Goods in St. Joseph, Michigan. Injection-molded decorative plastics and vacuum-formed, extruded and rotational cast plastic products were being produced in two plants in Evansville, Indiana, by Kent Plastics, another Ball subsidiary. A Kent subsidiary, Kent Plastics United Kingdom, Limited, was operating in Enniskillen, Northern Ireland, supplying decorative plastics to United Kingdom and European markets.

A hopeful subsidiary project had been launched in Boulder and Muncie, the Pantek Corporation, offering a unique housing system involving the use of structural panels.

In far-away Huenxe, West Germany, Ball-Grillo Micrometal GMBH and Ball-Grillo AG, Ball affiliates in the metal and chemical division, were producing photo-engraving plates for the

European market, with a sales office in Zug, Switzerland.

As if this complicated picture were not enough for a new president to contemplate, there were other influences, beyond corporate control, which had to be confronted. There was an economic recession during 1970 which, while weathered successfully by Ball, lowered earnings below expectations, even though net sales of $171 million were up 6 percent over 1969 sales figures of $161 million. Net income of $6 million was second best in the entire history of the company, but that followed a 1969 record of $7 million.

There were urgent external pressures to consider. Environmental concern was touching practically every industry, and, of course, Ball, with its wide range of products, was in the thick of a virtually obligatory program to ensure the environmental compatibility of all of its products and processes. As part of that program, the corporation was repurchasing used containers at all glass plants, acting in concert with an industry-wide effort to promote glass as an ecologically acceptable, recyclable material.

In addition to the rising environmental concern and the generally unfavorable economic conditions of that year, Ball faced record high interest rates, inflationary cost increases, and lengthy strikes in the trucking industry and at plants of several major customers. Start-up costs at the new Greeneville zinc plant were substantial, and there were costly delays while the three older zinc plants continued to operate during the breaking-in period at Greeneville.

In 1970, also, in spite of higher interest rates, Ball entered into a major refinancing program, completed in January, 1971, which resulted in bank and insurance company commitments of

$22 million.

Out in Boulder, Colorado, Ball Brothers Research Corporation (BBRC) was going ahead with the last Orbiting Solar Observatory (OSO) satellite under contract with NASA, to be launched in the first quarter of 1971. Subcontract work on "Skylab," an orbiting space workshop scheduled for launching in 1973, was also in progress. Meanwhile, Ball continued to march forward in the manufacture of innovative experimental space hardware, to exacting specifications, for government and scientific agencies, both domestic and foreign. Quite a stride beyond the company's first modest experimental interest in 1956, when Ball looked somewhat dubiously into the purchase of a small Boulder company manufacturing an electronic weighing device. The device proved to be a disappointment, but launched the fruit jar company into space as it led to contracts in government and scientific research beyond anyone's ability to foresee.

John Fisher, in Muncie, in 1970, found himself in charge of a diversified company with the promise of a great future.

The challenge of operating a company composed of all these segments, however, never became so compelling as to obscure in John's eyes his primary emphasis on people and ideas as the stuff of true accomplishment. Back in Muncie, he had gathered around him a team he was to put to severe tests in the next few years.

Most valuable, of course, were the input and guidance given by Edmund F. Ball, chairman of the Executive Committee and former president and chairman of the board, and Alexander M. Bracken, then chairman of the board and general counsel. Other company directors included Burnham B. Holmes, vice president of planning and corporate development, and R. Arthur Gaiser, corporate vice president of operations. Robert H. Mohlman was vice president in charge of finance and administration. Other directors were: John P. Collett; William M. Ellinghaus; Karl Nelson; Alvin M. Owsley, Jr.; Reed D. Voran; and Arthur M. Weimer. Vern C. Schranz served as corporate secretary as well as director of public relations. Company officials included: G. Robert Baer, group vice president, eastern divisions; William C. Hannah, group vice president, glass containers; Ruel C. Mercure, Jr., group vice president, western divisions; Kenneth M. Hay, vice president, commerical glass container sales; Ralph L. Hoover, vice president, industrial and employee relations; Patrick J. Kearns, vice president and general manager, Ball Metal & Chemical; Ronald A. Kibler, vice president, technical services; Herbert S. Kishbaugh, vice president, purchasing and traffic; R. Donald Bell, controller; William L. Peterson, treasurer; O. E. Bartoe, president, Ball Brothers Research Corporation; Roland P. Campbell, president, Ball Brothers Service Corporation; Harold W. Cochran, president, Caspers Tin Plate; Dan A. Gabrielson, president, Jeffco Manufacturing and Alpine Western; Robert H. Morehouse, chairman and chief executive officer, Kent Plastics; and Howard Oplinger, president, Ball Rubber and Plastics.

It was a far-flung operation demanding nothing less than a wide-angle lens viewpoint from the new president. But John Fisher had undergone a long apprenticeship for this moment.

As a boy in Walland, Tennessee, John had grown up in the shadow of the tanneries run by his father and grandfather, and he had always been fascinated by machines and manufacturing. When he attended the University of Tennessee as a member of the class of 1938, he majored in industrial management and even as an undergraduate worked in a local bank as well. He had a busy campus career, carved a name for himself in athletics and demonstrated leadership qualities in a number of extracurricular activities. Upon graduation, at a time when jobs were not in any great abundance, John was one of a team of four field secretaries, selected to visit the various chapters of his national fraternity, Delta Tau Delta. This became a useful bridge between academia and the business world for him.

John recalls, "It was a great lesson in learning how to live and relate to people, whether dealing with the dean of students, the president of the university, the president of the local alumni chapter, the undergraduate chapter adviser, the chapter treasurer or an ornery student. We had to be a combination auditor, public relations man, organizer, disciplinarian and diplomat . . . *and* we had to instill the principles of running a good business into those chapters."

On his tour around the country, John found some precious time to study manufacturing facilities and tour plants in a limited number of related industries. This was an experience which augmented the theory he had absorbed on campus in a way which few postgraduate studies could have offered. With this background, he was at first reluctant to give in to the challenges of a first cousin, Dean Malott, later president of Cornell University, erstwhile chancellor of Kansas University and then professor at the Harvard Business School, who "kept egging me on to go to Harvard Business School." Persuaded at last, John went East to obtain a master's degree in business administration.

While a student at Harvard, John did his

apprenticeship work at Ball Brothers during the summer vacation, learning the business of manufacturing from the ground up, when he often found himself using his muscle to load freight cars. (Little could he have guessed while at such lowly tasks that in 1980 he would serve as chairman of the National Association of Manufacturers, one of America's most prestigious honors.) In 1940 before entering graduate school, he married Janice Ball, the daughter of Edmund B., one of the five founding brothers. Two years later, upon graduating from Harvard, with several attractive job offers in hand, he was convinced by George A. Ball to work for the ailing fruit jar company. Envisioning a "great opportunity" to bring about needed changes in this fine old company, he returned to his apprenticeship and made the most of it, working under the watchful and experienced eye of founder Frank C. Ball, still president of the firm and fully in command of the situation though in his eighties.

Even at the top level, the ranks had been decimated by deaths and the demands of the war. John found himself brought in as a director of Ball in 1943 to help fill the board's depleted numbers, but he served well and has held the position ever since.

In those war years, Ball had all it could do to keep itself together in terms of management. Everybody was lending a hand. Meanwhile, Ed was away in the military. John corresponded with him frequently and saw him as much as possible during Ed's first tour of duty—at Middletown Air Depot, Olmstead Field, Pennsylvania, at Washington, D.C., and at Indian Town Gap, Pennsylvania, just before taking off for the combat zone in the European theater.

After that, with others of the family and man-

agement to be called into service and away from the enterprise for war duty, Ball Corporation in Muncie was on its own for the war's duration.

As we shall see, it was to engender some problems which almost threatened the existence of the firm. But for the time being, it was John Fisher's job to help keep things moving until the postwar years could bring the team back home.

John's personal career for the next 20 years immersed him in manufacturing, first in the glass division, then in paper, zinc, and rubber. Finally, he became vice president of all manufacturing. He had stints in every facet of the Ball operation, as general manager of a "sick division," nursing it back to health, before moving to vice president for sales, vice president of marketing and public relations and, later, manager of the glass container group.

Thus, it was an experienced and informed executive who took over as president in 1970. Even so, it was an awesome responsibility, and it must have given even a man of John's naturally optimistic and aggressive temperament some pause.

John grasped the helm firmly from the beginning. During his first year as president, the company's strategic planning program was re-evaluated, refined, and the following objectives outlined:

1.) To obtain a growth in earnings per share and a return on equity comparable to the performance of the top quarter of American business, which requires the corporation to increase earnings per share 10 percent or more each year, and to maintain a return on shareholders' equity of 15 percent or more per year.

2.) To develop a reputation in the financial and business community as an effectively managed, quality minded, technically oriented, aggressive growth corporation.

3.) To develop and maintain a reputation of outstanding citizenship in each community where the corporation operates.

John also noted, "This planning process involved gathering a strong fact base to determine the company's capacity for growth. Strategies were then charted for a course which would build on those areas presenting the best profit opportunities for long-range growth."

There was more than operational planning on the president's mind. The continuity and self-renewal at Ball had always been among the company's greatest strengths. He saw the company now as "doing it on our own." As Ed Ball gave the command to John Fisher, the corporation had come to a point where many opportunities were becoming available.

John determined to sail in new directions which had timely implications.

"We had been concentrating very heavily on our strategic and operational planning. The emphasis had been to perfect our data base, place ourselves in a ready position, know the facts about our customers, and ascertain the capabilities of our various production units so that we could match up these strengths and maximize our output," he said.

Concentrating on corporate strengths, John saw each facet of Ball capability leading up to new but related ventures. He now says: "We knew the packaging business pretty well. We'd been in glass, plastics and paper packaging and we had our closure business. The idea was to look at the packaging business and see if there were any new innovations on which we could capitalize. The new two-piece can was an obvious answer. We developed that idea and moved into the market in a timely and innovative fashion."

Three additional metal container facilities were constructed, one in Williamsburg, Virginia, in 1971, another in Findlay, Ohio, in 1972, and the most recent in Fairfield, California, in 1976. Throughout the decade since Jeffco was acquired in 1969, sales rose and profits kept pace with the challenge. Jeffco's sales figure at the time of purchase was $17 million. By 1979, the company's Metal Container Group reported yearly sales well above $200 million.

More was "coming on" during the early years of Fisher's presidency. There was expansion, too, in the rubber and plastics fields. A new plastics plant was built at Fort Smith, Arkansas, in 1974. The same year, the company acquired a rubber facility from Litton in Batavia, New York. Unimark Plastics, of Milroy, Pennsylvania, joined Ball in 1978.

Then evolved the nagging issue of whether the company should "go public." Although the firm had an ever-increasing number of both family and employee stockholders, it remained essentially a privately owned company. The 1960s and 1970s were years of important expansion through mergers and acquisitions for companies here and abroad, but Ball was partially stymied by its "private" status.

"We'd been frustrated in some of the acquisition possibilities," John says. "We passed up a number of opportunities because we didn't have a publicly owned common stock to offer."

The issue was debated at great length throughout the 1960s, but was set aside in favor of keeping with Ball's tradition as a private concern. By the early 1970s, enthusiasm for a public status grew and again officials agonized over various proposals and approaches. At long last the company "went public" on July 13, 1972, and its stock was traded "over the counter." Later, on December 17, 1973, its stock was traded on the New York Stock Exchange. The initials by which it would be commonly known in financial circles and listed on teletype machines relaying its market status around the globe: BLL.

After accomplishing this move, Ball ventured into more "exotic" concerns, although it was to remain essentially a packaging company. It continued to expand its space program, which has centered mainly on such areas as antennas, lubrication of hardware for space environment, cameras, telescopes, and data-recording instruments to study the sun, the moon and other planets. Many of those instruments and satellites sent into space, designed and built by Ball scientists, have provided data which now aid in predicting weather patterns, tracking ocean traffic, and monitoring pollution, crop diseases, and forest fires. A boost to this program came with the acquisition of Time-Zero Laboratories of Gardena, California, where additional sophisticated instrumentation was developed such as magnetometers, spectroheliometers, infra-red telescopes, and gamma ray detectors.

Under the Ball Brothers Research Corporation umbrella of technology came a growth thrust that would show Ball moving into areas of modern concern and future promise. BBRC ventured into the computer-related business, its strategy in this field being to produce ever-improving designs to make its products the best in two or three markets, always with an eye on future advancements. Here a modern version of Ball's old philosophy reigns: "If you don't obsolete your own equipment, the competition will."

Reflecting this philosophy, Ball has known success with several computer-related products, such as disc drives used to store and retrieve data. Data display monitors for some of this equipment come from the Electronic Display Division, located in Blaine, Minnesota. Concurrent with the company's rapid progress in space technology, another section of the research and development department was making significant contributions in improving existing product lines. The home-canning lid gasket, which had been manufactured for years at the rate of 60 lids per minute, was totally revamped, seriously researched, and today's rate is 600 lids per minute. Coatings to resist abrasion on glass surfaces were developed and their usage is now found throughout the glass container industry. An active patent department was started to protect the continual flow of novel developments coming from the R&D personnel, and today royalty income is an important item on Ball's income statement.

The demand for Ball kerosene cans manufactured in 19th century Buffalo never returned, but Ball Corporation, in a sense, re-entered the petroleum field with the acquisitions of Avery-Laurence, an operation based in Singapore, and the formation of Ball-Reid Engineers of Oklahoma City, Oklahoma, both engineering and managing concerns involved with petroleum processing facilities.

Again the company utilized its manufacturing expertise when it entered the agricultural field, but not as before in relation to home canning. It was more reminiscent of the time the Ball brothers went west to build a system to bring water to the residents of Pueblo, Colorado. Modern overseas irrigation operations began when Ball signed a contract with the Libyan government to supply the first of a series of irrigation projects designed to expand Libya's agriculture production through modern technology with devices such as center-pivot irrigation machines. These skills were later exercised when Ball aided irrigation projects in Saudi Arabia and also in the U.S. Southwest for Navajo Indians.

Ball purchased in 1974 the environmental-monitoring business of Sierra Research Corporation and integrated it into BBRC operations.

John describes these ventures as evidence of corporate sensitivity toward timely concerns in a changing world. Ball's first attention to the petroleum business came during the energy crunch of the early 1970s. When world hunger and self-sufficiency became vital issues, Ball improved methods of irrigating desert regions. The company's production of seamless cans came in response to such issues as pollution control, the conservation of valuable resources and the potential problem of lead poisoning from seams in cans. The call of the Computer Age was heeded early on and was followed by Ball's entry into computer markets.

But Ball maintained its base in the packaging field and continued to expand these operations in the 1970s. In the tradition of founder Frank C. Ball, whose glass-machine patents helped bring that manufacturing art into the 20th century, Ball Corporation now applied its technological know-how and introduced electronically controlled glass production. Technological improvements helped increase production output. In 1970 the glass container division represented 31 percent of total company sales, but with technological improvements and the company's competitive stance in the market, sales more than doubled in five years. By the mid-1970s, when the metal container division emerged as a strong facet of the

company, claiming 5 percent of the national market, more than half of total sales were due to its success in the packaging field. While John Fisher guided Ball's growth in these areas, he never lost sight of a goal he had set for himself when he first became president and chief executive officer in 1970.

"My entry on the scene at that time caused me to reassess our organization," he recalls. "I wanted to take a real good look at it. The first move I made was to start looking for a successor because I felt that the most important assignment of a chief executive officer is to be sure he has someone to back him up. For two years, I looked for that person; then I found him in our own organization."

The man he found was Richard M. Ringoen, vice president of the aerospace division. He was indeed uniquely fitted for the position.

Dick Ringoen had once considered being a minister, a banker, or a lawyer—quite a span of ambitions—when he was pumping gas for 10¢ an hour in his Iowa hometown of Iowa Falls at the age of 12 back in 1938. But his stars were to take him into other disciplines. He earned his B.S. with honors in electrical engineering at the State University of Iowa in 1947, and followed with an M.S. in electrical engineering at the same university in 1948.

Ringoen's zest for technical work was indicated by his work on a doctoral thesis project—measuring the moon's temperature with radio frequencies to determine its surface material characteristics. The Apollo program subsequently confirmed his findings. Something of the expertise indicated by this study is reflected in Ringoen's seven patents in communications, navigation, and electronic circuitry.

Before he arrived at Ball in Colorado, Dick spent eleven years with Martin-Marietta, where he was instrumental in the development of the Titan ICBM, back in the days when Russia and the United States were racing to see who could outdo the other in armament capability by developing the "ultimate weapon."

Prior to that, Dick had been with Collins Radio Company, where he worked on projects which were to singularly prepare him for his later years at BBRC.

John Fisher could find out such things about Ringoen from the record. But what John wanted was something less technical: What kind of man was Dick?

Today, in 1980, as we chat about these things in the Ball executive offices, John admits with a smile that he "checked him out pretty carefully. I even went to his hometown and talked with some of the people who knew him when he was a young man."

John found out a lot of things about his man which were compatible with the Ball tradition. Dick is a family man, and very active in religious and civic affairs. He is also an outdoors and fitness buff who still maintains a ski cabin in Vail, Colorado, for vacation schussing, all of which would recommend him to the tall man from Tennessee who still has the bearing of an athlete. Another plus for Ringoen was something he shared with research director Art Gaiser—in spite of his technical expertise, he has a flair for reducing technicalities to understandable terms.

And well it was that Dick Ringoen could translate and modify these technologies into everyday speech, and help change the public impression that Ball was only a company making fruit jars in Middletown, U.S.A.

John, half closing his eyes in the Muncie office and thinking out loud across the table from Dick, is speaking with utter objectivity and calm certainty.

He says: "As I looked into the future I felt that this company needed to increase its strength in the area of high technology and the application of sound engineering principles to our manufacturing business. My own strengths were more largely in the field of manufacturing. What we really needed to complete the team was a man with a very strong engineering background, who also had administrative capabilities. Dick certainly has demonstrated these qualities and also possesses the personality and strength of character to lead an organization through tough times, at the same time anticipating and providing for the difficult problems ahead for a business of this size. We've grown a lot together here. It didn't take me long to find Dick's strengths, what he could do best, and then to lay things in his lap when I felt he was ready for them and set to perform."

On his side of the conversation, Dick Ringoen responds in the orderly syntax of an engineer. He talks without a great deal of vocal emphasis, but what he says carries its own weight.

"My opportunity at Ball was unusual," he says. "Here was a family-held company which was so diversified, so well commanded, where management really didn't have to worry about takeover problems or loss of shareholder support. That added up to a strong organization which offered a great deal of freedom for an executive like myself who has always been interested in the exercise of positive management. Such an opportunity is very unusual in corporate management, and I could hardly imagine a better scenario of something to step into as president.

"So many companies are very political at higher levels, filled with all kinds of jealousies which threaten an individual's working environment. In that kind of atmosphere it's difficult to do a good job because you're always worrying about the wrong things. Here at Ball, I discovered that after we figured what was the right thing to do we could go out and do it. In so many corporate environments, you know that something is right, but you just can't do it because it's not possible under the existing interrelationships. At Ball, it *is* possible to a very large extent. That means we are able to accept our responsibilities in managing an organization, and to offer a genuine accountability."

With this attitude, Dick Ringoen acted promptly and vigorously after becoming vice president of operations in 1974. That he had John Fisher's confidence was evident when John said: "I think our relationship has been rather unique. Fortunately, we both understand that we don't have to do a lot of explaining about events to each other. We talk the same kind of language about people, and I think our manner of operating is also quite similar. We agree that when you delegate responsibility to others, you turn associates loose on the problem. Let them use their ingenuity, and if they need some help, they know they can always come back here and talk to us. I think that maybe the doors on our respective offices have been closed only a half dozen times since we've been in this building, and that reflects the open-door attitude that people have toward both of us. Dick operates the company in a manner in which I have complete confidence. I think the combination has brought about a good balance in managing this business."

One of his most effective moves was to rally together the people in the operating divisions (John and Dick share a belief in teamwork and communication). Dick instituted a system of visiting group managers around the country on a monthly schedule.

Dick placed more emphasis on operational performance, thus strengthening the line responsibilities. "Staff people have to be very supportive of line operations," he says. "Their mission is to service the line and not be a problem to them. They're not in a position of control, busy with building up their own empires or anything like that, but they really must help the line people accomplish their objectives."

John adds another point to Dick's analysis:

"Dick recognized where our strengths were in these staff groups and directed the operational people to call on those strengths. We emphasize, in our monthly review sessions with each of the operations, their performance against their plan. Incentive compensation is directly related to performance. So we really put some teeth in the planning program by affecting managers' pocketbooks in relation to their performance."

Another plan which Dick devised in his approach to bring the divisions of the company closer together was to hold his monthly meetings in various locations throughout the company so that one division could see how the others operate. In this manner, Dick formulated a cohesive group of operating units within the company. What his plan amounted to was similar to a board meeting in a small company, held every month, where policies can be examined or changed right then and there. "We make a lot of decisions on the spot at those meetings," he remarks. "Monthly, our Executive Committee follows those operational meetings. Thus I know what is going on out there in operations, and I can present their ideas at the top level almost immediately for Executive Committee discussion and, where necessary, approval."

It works. Says Dick: "In general I think our managers thrive in this environment. They stick with us because they are inspired by this challenge."

Acquisitions made in previous years had established beachheads in solid industries, mostly in packaging and related fields. These companies at the time of purchase had brought with them a fierce pride of ownership as well as name and trademark recognition in their fields. It would have been impractical to change their names at any earlier time, but, by the 1970s, their association with the parent company became an integral part of their identities. Dick saw the chance to rally the actual divisions around a central point— "Ball"—and moved boldly. All those organizations mentioned earlier in this chapter, like Jeffco Manufacturing, Kent Plastics and Caspers Tin Plate, were renamed to include the name "Ball" and the function of the operations, such as the Ball Metal Container Group instead of Jeffco Manufacturing, and Ball Metal Decorating Division rather than Caspers Tin Plate.

He was also interested in underlining the entrepreneurial character of the corporation: "We initiated some new ventures of low capital intensity where, if you went wrong, you wouldn't lose too much money. Some of them *have* gone wrong, but still they don't involve a lot. I think we were all trying not to be called a conglomerate, and really rally around this diversified packaging company image which emerged in the '70s."

This emerging "image" is Ball's profile at its century mark.

Today Ball is a diversified manufacturer with primary emphasis on packaging products for food and beverages. Other products and services are provided to industrial and high-technology markets. From its corporate headquarters in Muncie, the company directs the operations of 21 domestic and four overseas facilities and employs more than 8,500 persons.

Its principal business is packaging products, including glass containers for the commercial food industry and metal containers for brewers and soft-drink fillers. The company is a major producer of home food preservation products, including closures, for the consumer market. Ball also supplies metal lithography services, special coatings and equipment to packaging product customers.

Its packaging facilities include glass container plants in Asheville, North Carolina; Mundelein, Illinois; Okmulgee, Oklahoma; and El Monte, California; and a home canning closure plant in Muncie. Metal container plants are in Williamsburg, Virginia; Findlay, Ohio; Golden, Colorado; and Fairfield, California; with headquarters at the Ball Colorado Office Center in Westminster, Colorado. Metal decorating facilities are in Chicago, Illinois.

Industrial products include molded, extruded and vacuum-formed plastic products for the appliance, automotive, construction, medical and electronics industries. Fabricated zinc and lead products as well as shells and grids for batteries are manufactured. A metal service center supplies light-gauge steel and aluminum sheets and special cutting and slitting services.

Ball's industrial products facilities include plastics plants in Evansville, Indiana; Fort Smith, Arkansas; Milroy, Pennsylvania; and Enniskillen,

Northern Ireland. Rolled zinc and lead products and specialized chemicals are produced in Greeneville, Tennessee. The metal service center is located in Chicago. Petroleum equipment manufacturing operations are located in Indonesia.

Technical products include the development and manufacture of systems, subsystems and components for NASA, the military and international space programs as well as highly sophisticated instrumentation systems for space science experiments. Video monitors are manufactured and sold for commercial and industrial application. Advanced electro-optical products and various computer-related equipment are also manufactured. The company designs, procures, installs, and operates irrigation systems used in large scale agricultural developments, principally overseas.

Facilities in the technical products area are located in Boulder and Broomfield, Colorado; Blaine, Minnesota; Sunnyvale and Gardena, California. The productivity of the technical operations of the company may best be illustrated by its patent activities. As of this writing, 322 active patents have been assigned to Ball Corporation and 89 patent applications are on file.

In order to get a definite handle on this company, a neat package of words for a rapid understanding by laymen, Ball has been termed the aforementioned "packaging company with a high-technology base." It can also be said to be a manufacturer's manufacturer, or, in the words of John Fisher, an "industrial products business."

This starts John thinking about the future of the company:

"Our future course is to follow a strategic plan to continue to grow in the same percentages as the last five years. But these objectives become tougher all the time. As our base increases, our success penalizes us. Our challenge is to use each dollar more effectively than we have in the past, to maintain margin improvement and enter into some less capital-intensive enterprises."

John returns to his thoughts on ideas and people.

"When I think back to when I became president of the company, I see that our greatest accomplishment over these years has been in building a very solid organization. And when you have people, skilled people, in the right assignments, you can seize on opportunities."

Dick Ringoen, the man who stresses technology and teamwork, agrees. In his own words: "We have both the advantages of family heritage and the vitality to move with the times. As for the future, I would like to see the rewards to our shareholders continue to grow and our employees rewarded while feeling they have had opportunities to fully utilize their talents. A broader objective is for Ball to be a significant force in preserving our economic way of life in this country."

What then of Ball at the century mark, of satellites in space and agricultural efforts in arid lands halfway across the globe? Or data display terminals used by computer operators throughout the world?

Messrs. Fisher and Ringoen could sum it all up by saying: Follow the humble fruit jar from the days of home canning in mother's kitchen to a point where the commercial container market arrived. Add to this the influence and technical know-how of the space and computer ages, mixed in with a large management gift for entrepreneurial expertise, and you come up with a good recipe today for Ball Corporation. But today is only the eve of a second century of achievement.

The first century was prologue.

Part One

Ball Today: People and Ideas

Ball Corporation, according to Chairman John W. Fisher, has an
important investment in "people and ideas." The people are
those diligent employees who have moved
the company into a position of corporate leadership. Their "ideas"
are those new approaches to diverse products: from space and computer
equipment to seamless beer cans to maintenance-free battery grids.
The success of this combination is seen in such accomplishments as
membership in the select Fortune "500," with sales figures now
above the half-billion dollar mark. The company has distinguished
itself as a volume supplier of industrial products. According to
Mr. Fisher, "We seem to have a propensity to take on an opportunity
and then know what to do with it. . . . We make and sell large
quantities of an item. That's what we seem to do best."
Illustrated here, these products and the people and ideas
behind them show the company in a strong position at its century mark.

Directors of the Corporation gathered in Muncie for a meeting on
January 22, 1980. Pictured are:
Seated-(left to right) John P. Collet (Honorary Director),
Betty M. McFadden, Richard M. Ringoen, John W. Fisher (Chairman),
Edmund F. Ball (Honorary Director and Chairman, Executive
Committee), and Alexander M. Bracken
(Honorary Director and Chairman, Pension Committee).
Standing-(left to right) Richard M. Gillett, Burnham B. Holmes,
Alvin M. Owsley, Jr., Robert M. Spire, Delbert C. Staley,
Reed D. Voran, R. Arthur Gaiser, Robert H. Mohlman
and Arthur C. Weimer.

Ball Corporation, in the words of John W. Fisher, is a "packaging company
with a high-technology base." Today the company maintains four metal
and four glass container facilities, geographically dispersed, which
account for more than 50 percent of its total sales. But Ball is not a firm
to forsake its heritage. Slightly to the right and center in the
above photograph is a Ball fruit jar, which the company continues to
manufacture to this day. Home canning, once a necessity in the American
kitchen, is still popular. It is a classic reminder of the French adage
that "the more things change, the more they remain the same."

At right is the Ball-built camera for the Viking Orbiter which relayed pictures of the Martian surface (inset) to earth-bound scientists. Besides building cameras, Ball is also a leader in the making of lubricants, antennas and telescopes for spacecraft.

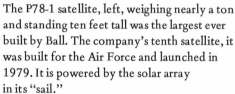

The P78-1 satellite, left, weighing nearly a ton and standing ten feet tall was the largest ever built by Ball. The company's tenth satellite, it was built for the Air Force and launched in 1979. It is powered by the solar array in its "sail."

Nineteenth-century French science writer Amedee Guillemin once asked, rhetorically, "What is the sun? If [science] could solve this . . . it would be nearly capable of solving the entire universe." More than a century later, scientists continue their relentless study of the burning orb, assisted in a large way by Ball Corporation, which has devoted much of its aerospace program to solar scrutiny. Perhaps its most dramatic contribution came in its work on Skylab, launched in 1973. At right is a view of the sun taken from Skylab by a Ball-built camera. The solar flare seen here (and called, quite poetically, "a great prominence" in scientific language) was a rare moment in solar study. The eruption stretched 300,000 miles high, making it one of the largest in a decade.

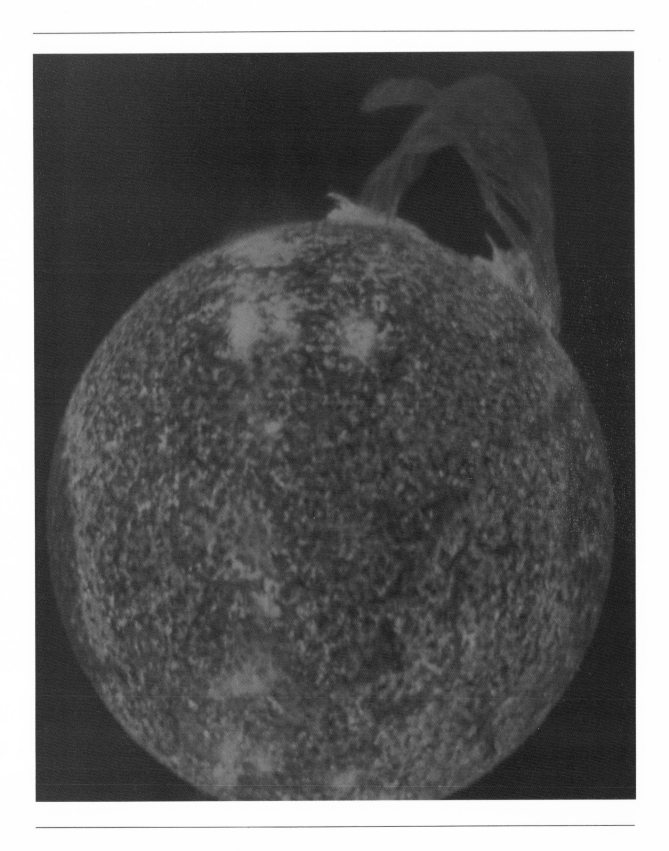

BALL TODAY: "PEOPLE AND IDEAS"

Ball's expertise in technical products extends over a period of more than 20 years. With the acquisition of Miratel in Minnesota, Ball entered into manufacturing industrial and professional television displays which can be found in broadcast stations, airports and in visual display terminals used in modern newspaper offices. Ball's Electronic Display Division is now the leading independent producer of data display monitors in the world.

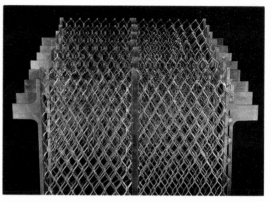

The Greeneville, Tennessee, zinc plant was put to use to manufacture this lead grid for a maintenance-free battery. The Greeneville plant is the largest facility for the continuous casting of non-ferrous metals in the world.

Harkening back to the days of Ball's manufacture of rubber sealing rings on fruit jars and following the company's journey to diversification after that point, the company observer discovers Ball producing plastic cups today, a further lesson in the evolution of this firm's history. Having been in the rubber business, it was almost an inevitability that Ball would produce a plastic container. Other expansion in the plastics field has helped the Industrial Products Group to grow. Ball's plastic molding operation now services medical, electronic and specialized industries. It acquired Unimark Plastics in 1978.

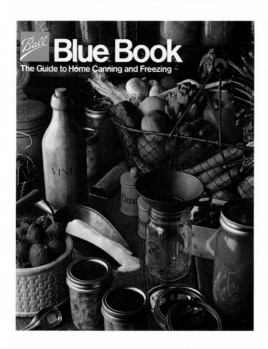

The *Ball Blue Book*, left, is an international symbol of the company's prominence in the world of food preservation. The concept of a book detailing the fine art of home canning was first conceived by George A. Ball, who published *The Correct Method of Preserving Fruit* in 1909. At left is the book's milestone 30th revision. The company continued an international presence in this broad field when it made the Libyan Sahara bloom, below, by using a central-pivot irrigation system, just an example of Ball's agricultural endeavors made possible by the application of technological advances.

A view in the world's only fruit jar museum, located just off the main
lobby in the Muncie office building. The jars serve as rainbowesque reminders
of Ball's rich heritage.

Chapter 2

RECOLLECTIONS

by Edmund F. Ball

They were gone many years before I was born, my grandfather and grandmother, but I can recollect hearing so many things about them I feel that I must have known them.

They both came of early pioneer stock, which had migrated from England to America in the early 1600s—first to Connecticut, then to New Jersey; served in the Revolutionary War; then moved to Vermont, and crossed the Canadian border, eventually settled for a time near Ascot and Stanstead, west of Toronto. The two families, the Balls and Binghams, followed strangely identical courses, but it wasn't until around 1842 that they eventually met in Trumbull County, Ohio, where they homesteaded.

In 1834, William Ball, his wife Marcy Harvey, and their nine children moved from their farm near Ascot by way of Buffalo to a location north of Warren to clear a farm out of the wilderness. Numerous families with roots in the United States migrated from Canada during the decade of 1830 to 1840 because of economic and political unrest, church-related problems, and the feeling that in the States lay greater security and opportunities for the future. For whatever reason, the important thing from the standpoint of this chronicle is that two of the sons of William and Marcy would play indispensable parts in the lives of the Ball

brothers. Lucius Styles, the eldest son, would become the father of the Ball brothers; a younger son, George Harvey, later a distinguished Baptist minister in Buffalo, would encourage and support his young nephews in their early business ventures. He would advance the $200 with which they would purchase the "Wooden Jacket Can Company," the little business which 100 years later would be known as "Ball Corporation."

The Binghams with their daughter Maria and her six brothers and sisters made the almost identical migration in 1840. Maria would become the mother of the Ball brothers.

As I have heard him described, the oldest son of William and Marcy must have been a remarkable person of many talents. In his early twenties, he left the family's farm for New Orleans and spent a couple of years there working as a carpenter and builder. During his return trip, while traveling through Kentucky, he learned of certain valuable oil and gas producing properties that he was able to promote and develop a number of years later.

From family records, photographs, and descriptions, he was a striking individual—tall, straight, black-haired and bearded, long bushy eyelashes, deep-set black eyes that twinkled with cheerfulness and good nature. He was known by his friends and neighbors respectfully as "Squire"

Ball. His judgment and fairness were so well considered in the area that, in those days before an abundance of lawyers and the availability of law courts, he was frequently called upon to settle disputes; his decisions were generally accepted as final.

As a farmer, he experimented with various types of agricultural products. He invented an automatic horse-drawn hay rake that pre-dated the development of many modern farm implements. He built a mechanical hay baler which he and his sons used to bale hay to take to the Cleveland market to sell. He invented and patented a barrel for shipping eggs that was a prototype of containers used for that purpose today.

Shortly after his return from New Orleans, he met, courted, and married an attractive and vigorous young schoolteacher, Maria P. Bingham. They acquired an improved farm near the town of Greensburg (which no longer exists) in Trumbull County, near Warren, Ohio, and there the five Ball brothers and their two sisters were born: Lucina in 1847, Lucius in 1850, William C. in 1852, Edmund B. (my father) in 1855, Frank C. in 1857, Frances in 1860, and George A. in 1862.

The third son, Frank C. Ball, in his Memoirs written in the middle '30s, describes his recollections of his mother:

"She was a remarkably strong woman physically and mentally with a sweet Christian spirit. Her home and her children were all and all to her. Her chief ambition was to bring up her children with good, strong minds and bodies with right ideals and true Christian principles. She urged us to always stick together and help each other; it was through her influence and her advice that we five brothers have been associated in business throughout life."

Sometime after the death of her husband in 1878, she came to Muncie to live with her sons and died there in 1892. The influence of this fine woman remained with them and profoundly affected their personal philosophies as businessmen and citizens.

Early in life the children learned by the example of their parents that benefits came from hard work and adherence to the principles of honesty, compassion for and understanding of their fellow men, and mutual respect. All of the children inherited many of the talents, characteristics, and high ideals of their parents.

Many years ago my cousins, Arthur and William, and I accompanied my uncles, Frank and George, to the lovely little town of Canandaigua in upper New York State. It hadn't changed so very much from the way they remembered it to be when they lived there as boys. Together, we visited the beautiful cemetery on the upper banks of "Sucker Brook" where they used to play and trap muskrats. We located the graves of these two fine people, their father and mother, Lucius and Maria Ball, buried beside sister Lucina and little brother Clinton, who had died when only two years old. My uncles were deeply moved, as was I who watched them bow their heads in silent and reverent memory.

It is appropriate, I think, at this point to quote a sentence from one of Uncle Frank's eloquent speeches made some years later:

"I hold that from the home around which clusters the fond recollections of a mother, from the home around which we gather our family circle, comes the inspiration of hope, the courage by which men perform the duties of life; and the man who has not been thus inspired, who has lived for the applause of the world rather than the love

and esteem of his home and family, has missed much that this life was intended to convey."

It clearly expresses a philosophy in which the Balls believed and which they practiced.

At the time of the Civil War, the father was too old and the boys much too young to participate. Without such instantaneous communication as radio and television, the war must have seemed very far away. Nevertheless, "Squire" Ball almost got in trouble one time at the outset of the war when he took a herd of horses to sell in Kentucky, at that time a borderline state without clearly defined loyalties. Many of these horses belonged to neighbors who had entrusted him with the mission of selling them. In Lexington he was accused of being a Northern spy and all the horses were confiscated. Returning to his home with neither horses nor money, his neighbors threatened to carry off his farm machinery, grain, and tools in payment. But his wife, Maria, persuaded them not to do so and eventually all the money was paid back to his neighbors.

In 1863, in the midst of this awful war, seeking a healthier climate for the ailing father, they decided to move. Uncle George, by now a Baptist preacher of some renown in Buffalo, urged them to leave Ohio for what he thought might be the more bracing environment of a farm he owned on Grand Island in the middle of the Niagara River just above the Falls. In their migration a wagon bore the household goods, Mother, and the smaller children. Father and the older boys rode horseback and drove the livestock. The trip lasted several days, passing along the then-clean waters and lovely shores of Lake Erie. They camped at night, ate by campfire, swam, skipped stones in the water, and picked up shells from the beaches. It was exciting and fun for the children, who remem-

bered it fondly all their lives; but it must have been a trying experience for a concerned mother with small children and an ailing husband.

They spent two happy years on Grand Island. By then the oldest child, Lucina, was sixteen and the youngest, George, was only one. Even though they had been all quite young at the time, often when they were together in later years they would reminisce about the fun times they had on Grand Island. They would remind each other about an exciting incident like when they were in a rowboat, caught in the fast current of the river and almost swept over the Falls before one of them could grasp hold of a clump of weeds growing on a rock in the fast water, thus turning the bow of the boat toward the shore so they could row to safety. The older boys remembered one winter when their father went to Buffalo on business. He couldn't get back for three weeks because the solid ice over which he had crossed to the mainland had broken up and the floating ice made it too hazardous for the ferry to run. In his absence, the hired man decided to go on a drunken spree, leaving the boys to attend to the stock and do all the chores—heavy responsibility for boys not yet in their teens, but they enjoyed it. Sometimes they would reminisce about characters on the island, and when I visited there many years later with my Uncle Frank, he was disappointed not to find some of them still around.

That memorable visit to which I have previously alluded was made in the early '30s. Uncle Frank and Uncle George, then the only living brothers, with my cousins, Arthur and William, and me, revisited by automobile some of the scenes of their childhood. We caught the train from Muncie to Buffalo where Guy Dickason—long-time foreman of the garage, confidant, and part-time family chauffeur—met us at the station in a huge twelve-cylinder Packard touring car which he had driven on ahead. Guy was a fine mechanic, a good driver, and enjoyed the confidence of all the family from the youngest to the oldest. If he were living today, he could well be writing these recollections instead of me.

After breakfast at the hotel, we drove around Buffalo, saw Uncle George's Niagara Baptist Church (which is still standing) and 55 Main Street, the big warehouse where the two brothers, Ed and Frank, at one time made wood-jacketed containers. Here they had proudly painted the name "Ball Brothers" on the sign beneath the street number, leading to humorous queries as to who these 55 Ball Brothers might be. We drove around the Falls, across the bridge to the Canadian side, and onto Grand Island. There to our surprise and pleasure we found the old home in which they had lived still standing. It was quite large, with a hallway in the middle from the front door to the back. It was made of logs covered with clapboard siding and painted white. They remembered it as being cool in the summer and warm in the winter because of the insulation of these thick walls.

We knocked at the door, but no one was at home. We walked around the yard and Uncle Frank, having been seven years old when they lived there, found several familiar landmarks. Uncle George had been only a baby then, so his recollections were mostly from hearsay. Uncle Frank recalled several experiences, like the time they watched helplessly from the shore while ice fishermen on broken ice floes (crying out for help which no one could give them) were swept to their deaths over the Falls. Then there was their hero, the boy who had saved the life of his father when he became incapacitated in the middle of the river and could no longer row their boat out of the swift current. We drove around the island and Uncle Frank was clearly disappointed when he could find no familiar names on the mailboxes.

The family left the island in 1865, crossing the river to Tonawanda, where they lived for two years. While there the father became a partner in a grocery store which turned out to be a losing proposition. Money for the investment came from promotion of his Kentucky oil leases. Their home in Tonawanda was on the banks of the river and Uncle Frank recollected that with a long fishing pole they could have fished right out the front window. Here they caught and pulled out of the river logs that had broken away from lumber rafts far upstream and sold them back to the lumber mills for 25¢ each, as well as doing all sorts of odd jobs to augment a meager family income.

It was a relatively short distance from Tonawanda to the small community of Canandaigua where they spent their most formative years. Lucius, the older son, then 17, was assuming more responsibility as head of the family as their father's health continued to deteriorate. After their father's death he continued in this capacity until his brothers and sisters were educated and well established in their careers. Then at the age of 40 he fulfilled a lifetime ambition to study medicine at the University of Buffalo and became a practicing physician.

Canandaigua is now, as it was then, a lovely community located at the head of Lake Canandaigua. It was here that the family spent ten happy years. Here the children received excellent educations at Canandaigua Academy where the curriculum stressed the importance of learning the basic

"3 R's" and "declamation"—memorizing and reciting poems, scriptures from the Bible, and orations and writings of great statesmen and scholars. The brothers and their sisters, as well as the cousins that I knew, were all articulate and interesting conversationalists. I credit much of those abilities to their early training and the education that they got while in Canandaigua. Their boyhood experiences were not too different, I'm sure, from others in those days. They never considered their difficulties as hardships, but as challenges.

Long after they had become well-to-do, they recalled those days only as wonderful experiences. Today such circumstances might be considered underprivileged and their economic condition substandard. In an early business venture, $50 earned from a lemonade and fruit stand paid half a year's interest on the mortgage on the farm. It was the largest amount of money that they had ever made up until that time, but values were different in those days.

In the winter of 1877, my father had gone with a cousin, Charlie Barker, to work as a lumberjack in northern Michigan. Lucius, Will, and George remained on the farm with their parents and sister. Brother Frank was working in a paper mill at Shelby Center near Buffalo. Frank had left with his father's admonition: "Go, Frank, and see what the business is [His Uncle George had learned of a business opportunity in Buffalo that he felt was worth investigation]; I have every confidence in you and believe you will succeed." These were his father's last words to him since he died before Frank could return.

On occasion my father would tell about his experiences in the north Michigan lumber camps. One I well remember must have happened just before his father's death. It would have been the Christmas of 1877. All the other members of the lumber camp had gone to town to celebrate Christmas while Father elected to stay in camp alone. He was in the cookhouse when a terrible wind storm blew a huge tree down on the bunkhouse. It crushed the roof and fell across his bunk which was on the upper tier. Had he been in it he would have been seriously injured or perhaps killed. When he heard the teams coming, bringing back the other lumberjacks from town, he hid in the woods. As soon as they saw the condition of the cabin they began calling his name, thinking that he must surely have been in the cabin and might be lying there hurt, unconscious, or perhaps even dead. For a few moments he didn't answer them but then, realizing their real concern and that it wasn't a very good joke, he stepped out (much to their relief), evidencing that he was both alive and well.

It must have been only a few days after the incident that he got a message from Canandaigua that his father was dying and he should come home at once. He left camp, taking with him only the clothes on his back and a small amount of money, just enough for the railroad ticket back to Canandaigua. The trip took two days. It was snowing and bitter cold. On the train he and two or three other passengers were huddled around a hot pot-bellied coal-burning stove located in the baggage car. His heavy overcoat, a mackintosh, caught fire and it was only with some difficulty that they were able to extinguish it but not before the coat was completely destroyed. In weather still bitter cold and snowing, he arrived almost penniless and without even an overcoat. He borrowed a lap robe from the driver of the conveyance that took him from the railroad station to his home and wrapped it around himself to keep from freezing. His brother Frank had arrived shortly before from Shelby Center to join a grieving family circle now broken. Neither had arrived in time to see their father still alive.

While Lucius was the titular head of the family by reason of his being the oldest, Frank and Ed seemed to be more aggressive and the real decision makers. Father decided that he would not return to the lumber camps but would go to their Uncle George's home in Buffalo with brother Frank and try to develop some kind of a business there. It was at this time that the brothers resolved to heed their mother's advice and wishes that whatever they eventually did, they would do together.

It was 1878 and the children by now were reaching maturity. The oldest daughter, Lucina, was an accepted leader in the field of education, first as secretary of the Pratt Institute in New York City, and then as secretary in charge of curriculum of the newly created Drexel Institute in Philadelphia. Lucius, at the age of 28, was operating the farm, taking care of his mother and his younger brothers and sister, and looking forward to the time when he could begin the study of medicine. William was 26, helping his brother on the farm; so was the younger brother, George, then 16. Frances was 18. She was shortly to leave for Hillsdale College in Michigan. Ed, 23, and Frank, 21, were living with their Uncle George in Buffalo and making their first ventures into the manufacturing business.

For a number of years after the glass factory was moved from Buffalo to Muncie in 1887 (in fact, until his marriage in 1903), my father lived in the old Kirby House Hotel located on Main Street.

Mother was born in Terre Haute, Indiana, but lived most of her life in the East and abroad. Through mutual friends, she became acquainted with the Brady sisters in Muncie and it was while visiting Elizabeth, the oldest sister, who by that time was the wife of Frank C. Ball, that she met the "confirmed" bachelor who was to be my father. She was the daughter of a highly regarded Universalist minister who held several pastorates in the East and for a while in Scotland. To augment his rather meager income as a preacher, his wife, Adelia Swift Crosley, made the arrangements for and took travel parties abroad to Europe, North Africa, even Japan and India, long before the days of formalized commerical travel agencies and tour directors. When health would no longer permit her to travel, my mother took over these responsibilities on several occasions. It was quite an undertaking for a young lady just graduated from Vassar College.

On her last such trip the "confirmed" bachelor she had met at Frank Ball's made a special trip to England to see her. She always felt it was for the purpose of proposing. Apparently he lost his nerve, for he returned to the States without having done so. Nevertheless, later he must have mustered his courage sufficiently because they were married in 1903 in Indianapolis. Newspaper accounts indicate that it was quite a social event, with all of his brothers acting as groomsmen. For the wedding trip, instead of going to the usual fashionable resorts, they went west to visit such rugged places as gold mines in Colorado and Indian reservations in New Mexico and Arizona. Perhaps this accounts somewhat for my eventual interest in that part of the country.

With my parents' late start at raising a family, it is obvious that I was one of the later arrivals to join the Muncie clan of Ball. I was born in a double house that used to stand at the corner of Washington and Elm Streets. It is now the parking lot for

Muncie's leading funeral director, so it seems quite probable that I will conclude my career at the very same spot I began back in 1905. It was two years later, in 1907, that Father joined his brothers by building a home between William's and Frank's on Minnetrista Boulevard. The residence, now under the control of Ball State University, is known as the Minnetrista Center for Non-Traditional Studies. I am sure that my mother and father would heartily approve of this use of their former home.

My earliest recollections of the company's office are of rainbow reflections of sunshine filtered through the prisms of lead glass windows and falling on the brightly designed tile floor of the old office that used to stand on 9th Street. Next, I remember trips through the factory with Father, the heat and frightening noise of the blowing rooms where the glass jars were made. Then the comparative quiet and the coolness of the sorting department with the musical clatter of glass clinking against glass as sorters carefully selected fruit jars from the gleaming rivers flowing endlessly from the annealing ovens for transfer to the packing rooms at the back of the plant to be capped and placed in paper cartons for shipment or storage.

The brothers were fairly well advanced in their careers when I begin to recall them clearly.

Uncle Will, although quite active and athletic as a younger man, must have been a good deal like his father. He was, as I remember him, rather frail and in declining health. He had the reputation for being a tremendously effective salesman. He served the company as salesman and its secretary until his death at the age of 69. While a student at Canandaigua Academy, he had learned to recite numerous poems with great eloquence and feel-ing. As I think of him, in my mind's eye I can still see Horatio battling at the bridge, feel Lord Ullin's anguish at the loss of his daughter, and thrill to the "Charge of the Light Brigade."

Uncle Lucius, the doctor, was a quiet man, thoughtful, compassionate, with a shy sense of humor that endeared him to his family and friends. In addition to his private medical practice, he served as medical adviser for the Western Reserve Life Insurance Company which was then located in Muncie. I perhaps remember him best as being the uncle who always seemed to have time to take us fishing and the one who was called upon to bind up our wounds and abrasions and tend to us when we were sick.

My father was deeply respected and admired by folks in all walks of life. I remember one time when he came home late for supper (as he often did, much to my mother's despair). He was obviously quite pleased. He told of being asked to come down to Number 2 Plant Packing Room to meet with a group of employees who had assembled there. Since it was not explained to him why, he was fearful that some serious problem had developed and approached the group with considerable apprehension. When everything quieted down, a spokesman for the employees presented him with a gold pocket watch in appreciation for all the things that he had done for them. It became one of his most cherished possessions.

For many years, he served as treasurer of the little Industry United Brethren Church located near the factory. Each Sunday he would meticulously count the pennies, nickels, and dimes from the collection, and deposit the funds to the church's account in the bank, always adding to it a bit from his own personal resources. In a quiet way, he was active in many community affairs, although he preferred to stay in the background. Most frequently his name would appear as chairman of the building committee for such projects as buildings on the Ball State campus, the Masonic Temple, Y.M.C.A. and other community undertakings. His main interests were in construction.

It was at his wish that the Ball Brothers Foundation was organized and from his estate came its original funding. The foundation's first major project, Ball Memorial Hospital, completed four years after his death, was the culmination of one of his plans for the community.

Father suffered a stroke in the fall of 1924. I was 19 and a freshman in college. He died the following March leaving the entire community in mourning. Out of respect, factories, banks, and stores all closed at the time of his funeral. I was barely 20, quite young to be burdened with so many heavy responsibilities. The solemnity of Father's funeral service and the tributes paid to him from so many sources made deep impressions on my mind that have remained with me all my life.

Frank C., the fourth son, and my father were the closest of the brothers and shared many common experiences. When they were quite young, they were chopping silage in the barn on the farm. Frank was feeding the corn stalks as Ed chopped them up with a long blade of some kind. Frank got his right hand a little too far forward and my father, unintentionally of course, became the executioner of the first joint of his brother's index finger. This mutilation became a distinguishing feature which he used most effectively. Because of it, he learned to write a beautiful left-hand script. The flourish with which he affixed his signature to a document, with the paper turned sideways to accommodate his left hand, was a

sight to behold and long to be remembered. On occasion he could use that stub of a finger to make a point or to hold a person's or an audience's attention as effectively as some biologist might hold a beetle on the end of a pin for examination. His piercing eyes, magnified by the aspheric lenticular spectacles he wore after cataract surgery later in life, and his emphatic finger made him a formidable and effective discussion leader. It was under his direction that many civic projects were begun and successfully completed. He was a discriminating connoisseur of art. His collection left to Ball State University became the nucleus for one of the finest art galleries in the Midwest.

Having no sons of his own, Uncle George, the youngest of the brothers, treated me as if I were his. His thoughtful counsel guided me through many difficult situations. He had a quiet but persuasive way about him. Often a series of penetrating questions would lead to well-thought-out conclusions. Sometimes exasperated by the difficult questions he posed, those attempting to respond felt that he would have done well as a lawyer and prosecuting attorney.

Many stories are told of his shrewdness and his penetrating inquisitiveness. These experiences were not without their frustrations, but more often they would result in a more careful consideration of a problem, and eventually lead to the best solution. He served on the board of directors of numerous companies, educational and health care institutions. Notable among them were Borg-Warner Corporation, Nickel Plate Railroad, various banks, Indiana University, Ball State Teachers College, and Ball Memorial Hospital, each of which he served as director and the latter three as board chairman. He was catapulted into further national prominence in

1935 when, following the unexpected deaths of the Van Sweringen brothers of Cleveland within a few months of each other, he became the owner of their railway empire. This fascinating episode is covered in some detail in a later chapter of this book. He was also involved in politics and for several years he was a Republican national committeeman from Indiana. History held a great fascination for him. He collected important historical documents and gave them to libraries for preservation and display. Also he was instrumental in the restoration of historic landmarks. After his 90th birthday he was occasionally asked about retirement. His prompt reply was always, "I never entertained the thought of retiring."

Frances, the youngest daughter, married one of her college professors, Dr. William J. Mauck, who later became the president of Hillsdale College in Michigan (as did their son Willfred). She maintained a deep interest in education all her life. After her husband's retirement, they traveled extensively. While on a trip she died unexpectedly in Cairo, Egypt, at the relatively young age of 54.

I am greatly indebted to both Uncle Frank and Uncle George for the way in which they took me into their confidence after my father's death. Although I could add very little in the way of judgment, I was always consulted and kept advised of important matters and decisions.

It is my great privilege to have known so many members of this interesting family and to have shared in its heritage. It is my hope that the relating of my recollections in the earlier pages of this history will provide means for better understanding and greater appreciation of the events that are recorded in the chapters that follow.

Chapter 3

A MODEST START

Hindsight, the wisdom most easily come by, tells us that the name of Ball figuratively took off into space from a pad located only one story above the ground, at a point in time one hundred years ago.

The occasion was the launching of a tiny business at 29 Hanover Street, Buffalo—"The Wooden Jacket Can Company"—by two of the aspiring young Ball brothers.

Their capital: 200 precious dollars borrowed from paternal Uncle George, the wherewithal to purchase a down-to-earth process for providing containers to purveyors of kerosene and paint.

Their experience: almost nil.

Business record: two previous failures (the first, making containers for the fish-packing industry; the other, a rug-cleaning establishment which they abandoned because it appeared an unseemly business, and besides, they preferred to manufacture something).

Present prospects: on a scale of one to ten, about two.

Confidence: unshakable.

Not too sanguine a situation to the casual investor, but the omens, as we may see so clearly now, were all in their favor. Their assets were intangibles, it is true, but those unseen values added up to a reality: potential.

To begin with, they found themselves strengthened by a unity their mother had so often urged upon them, and which they would retain all their

long and productive lives.

What other men strive for, and most never achieve, was theirs already, nurtured by a family upbringing which blessed them with a philosophy in which hard work, spiritual dedication, respect for others, a nose for opportunity, and a zest for challenging the unknown were normal elements in their daily lives.

Added to that, they were the very personification of the unique American character which the French social philosopher, Alexis de Tocqueville, had discovered with such amazed delight during his visits to America in the 1830s. He could well have been talking about the Ball brothers in saying that "to an American no natural boundary seems to be set to the efforts of man, and in his eyes what is not yet done is only what he has not yet attempted to do." That kind of motivation came naturally to the brothers.

They were also in the right place at the right time.

America was recovering from the trauma of the Civil War toward the close of the nineteenth century and its citizens were in ferment in efforts to expand on all fronts—industrial, scientific, educational, political. The world began to notice the fledgling power, as railroads pushed into all of the settled parts of the country. George Westinghouse made it possible with the air brake. Alexander Graham Bell discovered how carbon granules would allow men to talk over a wire. George

Eastman invented the first successful roll film for cameras in 1880. The first safety razor was devised by Kampfe Brothers of New York City. Sarah Bernhardt made her debut on the New York stage. Longfellow, Whittier, and James Russell Lowell were the current best sellers in poetry. On November 2, 1880, James A. Garfield defeated Winfield S. Hancock for the U.S. Presidency. A nation of some 50 million (immigrants, sons of immigrants, former slaves, joined by the womanhood of America) was on the march.

And Buffalo, where the Ball boys were living, was, to hear the natives tell it, the center of this booming continent, a place which, in a somewhat grandiose self-image, was dubbed "The Queen City." In our time, we can appreciate that the city fathers were somewhat premature. But as they looked around them in 1880 perhaps it was not unreasonable to expect that Buffalo would never stop growing.

This bustling city on Lake Erie was athwart the long freshwater path to the West across the Great Lakes. It had eleven railroads connecting to the rest of the United States and the Erie Canal, a cornucopia through which funneled the myriad products of the New World to the east coast and on to European markets. In 1880 Buffalo ranked eleventh in the entire nation with $40 million annually in manufactured goods. It was to be the home of many great Americans. Two—Millard Fillmore and Grover Cleveland—made it to the White House. Mark Twain once invested in the local newspaper, *The Express*, and edited it himself, scattering shafts of wit among its pages reflecting the sparkle of blue lake waters only a block or so beyond the site of his labors.

There was simply no end to the heights which Buffalonians imagined for their city. Certainly, they did not lack for individual confidence. As one cracker barrel manufacturer said to his assistant, who suggested they take on another form of business as insurance against new methods of packaging: "Nonsense! If they make more crackers, we'll make the barrels to put 'em in!" Such heady intoxication of unchecked success could not have failed to infect the Ball brothers. Buffalo, as one could readily behold, was the right place to launch an infant business.

Yet, there was also the essence of rural America still lingering in the boisterous city, something which did not remove the Ball family too ruthlessly from the farm life which had been the chief pattern of their lives until now. They could hold fast to old faiths and old values without sacrificing them to their new existence.

Their Uncle George had never stopped trying to help. True, he had stumbled along with the boys in their ill-starred fish-kit and rug-cleaning efforts, but he was never one to take a step backward. Undeterred, he placed Ed as an apprentice with one of his parishioners, A.W. Aldrich, a local manufacturer of copper tea kettles and wood-jacketed oil cans, to learn the trade.

After Ed had been at work only a few weeks, his employer suggested a tempting opportunity. Mr. Aldrich, turned benefactor, would make the oil can division of the business available to Ed for an investment of only $200. This handsome sum would purchase some tinner's hand tools, a few boxes of tin, and several completed wood jackets. With this basic equipment, Mr. Aldrich predicted an enterprising young fellow could whip up a pretty good trade selling wood-jacketed oil cans to the paint and varnish dealers who in turn provided supplies to shipping chandlers who in their turn stocked the vessels riding at anchor on the

Buffalo waterfront. There was, "obviously," money to be made out there although the Aldrich method of producing the cans was patented and a royalty would be charged by patentee, Aldrich. The patentee also neglected to mention that his business had been losing money for some time.

The philanthropic Uncle George, nevertheless, put up the purchase price of $200 on loan to Ed. Never, in his most golden dreams, could he have guessed that here was, in embryo, a firm which a century later would reach sales exceeding a half-billion dollars.

Here was *the* start, this was *the* time, now in 1880.

Ed brought in brother Frank and they set to work with a will to make as many tin cans cased in wood as possible within the briefest time. As neophytes turned to the rough work of making wood jackets and attaching them to the cans, they were obliged to hire a tinsmith to perform the more skilled task of fashioning the cans themselves in one, two, three, five, and ten-gallon sizes.

Now a typical Ball reaction set in. The brothers had paid only one month's royalties to the wily Mr. Aldrich when they alertly perceived that the "patent" was simply in the peculiar way the jackets were put together and fastened to the cans. With a touch of their father's inventiveness they discovered that the jackets could be made much cheaper and better by simply lapping the ends and tacking them together with soldered tin, thus simultaneously doing away with a lot of unnecessary work and the royalty expense. With this skillfully applied improvement they could sell the cans at the established price, yet increase their profit.

Frank proved to be the master salesman. After

an opening sale of a dozen cans to some Buffalo paint and varnish dealers, and another direct transaction to some ship chandlers, Frank took to the road. He soon brought back orders from Cleveland, Toledo, and Detroit. Immediately, he and Ed brought in another tinsmith and two boys to help catch up with the new orders. Then Frank made another and more-extended trip to Cleveland, Toledo, Indianapolis, and St. Louis. He was so successful that the little firm was able to buy some used machinery and make plans for increased production.

Alas, just a few days after his return from St. Louis, their Hanover Street factory caught fire. Before it was under control it had completely destroyed all of their equipment and material.

Fortunately, they were fully insured and were paid promptly. They bought new machinery, rented another loft, at 55 Main Street, on the corner of Perry, and in a short time were again turning out salable cans.

To them, their new shop was unbelievably good fortune. Frank recalled in his Memoirs:

"Our new quarters were far superior to the ones that burned. The place had been used for manufacturing lanterns. The lofts (over four stores, three stories high) were connected by arches and stairways. We rented one of these lofts to begin with; the other lofts—twelve in number, each about thirty feet wide by three hundred feet long—were unoccupied, and as our business grew we rented additional lofts until we occupied the entire twelve.

"The stairway was located on the outside extending from the sidewalk to the office on the second floor. The office was quite pretentious. It was fitted up for a business much larger than ours and to us it seemed quite imposing. It certainly

was a great contrast to the dingy loft, with no office, formerly occupied by us, where a packing case answered for a desk."

By 1976 the same company was to move into corporate headquarters in Muncie, Indiana, costing in excess of $6 million, containing 116,500 square feet of such functional elegance that a panel consisting of members of the Institute of Business Designers, the American Institute of Architects, and the Association of Consulting Management Engineers would vote the headquarters the first award winner in *Administrative Management Magazine's* "Office of the Year" for 1976. The building also won the 1978 Man-Made Environment Award of the United Cerebral Palsy Association.

Yet, 55 Main Street, Buffalo, was to be the locus of an astonishing growth. For the next eight years it radiated an inexhaustible energy which converted into a prodigious flowering of new projects and plant expansion. So much so that when in 1886 the decision was made to open an office and build a glass plant in Muncie, Indiana, the local newspaper heralded the arrival of "one of the largest glass factories in the world."

But it was not offices and more space that were making the difference in 1880. For one thing, this family whose members meant so much to each other was now reunited for the first time since the death of the father. Brother Lucius left Canandaigua and joined them in the Main Street offices. George, who had been with a large hardware jobbing house as corresponding clerk, joined them. At that time Will managed to sell the old Canandaigua home and he and the mother moved to Buffalo. Lucina was with the famous Scribner publishing house in New York City, while Frances attended Hillsdale College. But the boys were together once more, and the presence of each strengthened the others.

They enjoyed their new camaraderie. Profiting from their experiences as farm boys, they bought a frisky young horse from a shipment which arrived in Buffalo from Canada. They called her Fanny and put her right to work. The sight of the little mare pulling the boys to work over the cobblestone streets of the city, as they "commuted" the two miles from home to work in their "democrat wagon," became a familiar one to Buffalo neighbors.

During the day, Fanny was also used to haul cans to the freight house. But she achieved a special kind of notoriety because of her nervousness and skittish fright over railroad trains. At the sight of a train or engine, Fanny would prance and sometimes even walk on her hind legs, right in the tracks, and earned the name "Ball's Circus Horse." She was not vicious, just a little nervous. The boys, with their farm background, had no trouble at all in managing her, and thought it great fun to see her perform when she was spooked by a train.

At this time an important evolution took place in their business. It was the process of conceptualizing and synthesizing certain policies and plans of action, probably entered into without conscious thought but as decisions made in the heat of competition. Yet these policies were to survive through the entire long life of the company as guiding principles. They were simple and direct. The company unfailingly moved swiftly to take advantage of any new advance in technology as soon as it appeared. The brothers sensed competition from afar and moved aggressively to meet it, usually with some innovative achievement which put them into a more commanding position.

The brothers' pattern of action and reaction was put to early use. The product line of small wood-jacketed cans was soon enlarged by the production of 60-gallon oil tanks suitable for grocery stores storing kerosene for retail customers. The big tanks were also sold to wholesale grocers and oil dealers, among them the Standard Oil Company. The new little business in Buffalo was getting into some very good company.

The perceptive Ball boys, perhaps remembering the storekeeping days, next cast an eye on their own one-gallon cans. These had originally been for the jobbing trade only. Now, with the addition of a spout and screw top to these smaller cans, and a sprucing up with a coat of varnish, here was suddenly a nice retail item.

That brought them a new challenge. Certain competitors were putting out a glass oil container. The Ball brothers, knowing that their own tin cans suffered corrosion from acids in the kerosene, quickly saw the handwriting on the wall. The risk-taking kinsmen lost no time in deciding to make their own glass oil can.

To do so, they imported the glass bodies from a plant in Poughkeepsie, New York. The Ball assembly line had about a dozen people running a ten-horsepower engine and boiler, with stamping press and dies for making the tin jackets, tops, and spouts. Their new product, however, displaced practically all of the older stock on the Buffalo market. Ball insistence on high quality and low prices inevitably gave them a large part of this flourishing segment of the oil business.

Their old enemy, fire, now tamed and harnessed, was put to work creating goods instead of consuming. Their innovation came about in the usual manner. The brothers couldn't help but note that the cost of bringing in their glass from outside (the expense of freight, handling of the packages, and plenty of breakage) was all over and above what they would be paying if the same glass could be made and fitted into the cans in one plant. So it was their good fortune when a group of glassblowers left the Poughkeepsie outfit following a dispute and started their own cooperative glass plant in nearby Olean, New York. Unnecessary to say, the Balls purchased their oil can bodies from the new group. Soon after, the Olean factory was destroyed by fire. Its alert manager, William Bryan, came down to Buffalo and emphasized the obvious. If the Balls would build a glass factory, he'd tell them how to go about it. They stood to gain from this expansion, in the long run, by eliminating all those shipping costs and breakage en route, while realizing some profit on the glass itself.

The Balls, never prone to quibble with the obvious, took the plunge. They bought a piece of land in East Buffalo, mortgaged it, and invested all their capital on hand in erecting their own glass plant and adjacent two-story brick building for the stamping works.

They were getting on. In 1881 the Buffalo City Directory listed the Ball business as "Manufacturers, Shipping Cans and Tinware." The next year, the listing read "Tin and Coopersmiths," a proud title somewhat lessened by the additional information that the proprietors, the Ball brothers, lived at 465 Niagara, with a small "b" attached to that address to advise that this was where they *boarded*. But, by 1885, it was "Ball Bros. Glass Mfg. Co. (F. C. Ball, pres.; G. A. Ball, sec. and treas.) 10 to 32 Porter."

Predictably, it was not long before the Balls were faced with the need to increase their glass-making capacities. Canadian contacts put them in touch with some glassblowers in Hamilton, Ontario, who had saved up their earnings and were anxious to invest it in the glass business.

They got together, and on February 13, 1886, the Ball Brothers Glass Manufacturing Company was incorporated officially under the laws of the state of New York. Capital was $50,000, divided into 500 shares of $100 par value each. Frank C. Ball was president; Edmund B. Ball, secretary and treasurer. Two Buffalo friends of the brothers, Frank R. Jones and Henry L. Jones, were stockholders. The remaining stockholders consisted of the glass men from Canada—Peter Menard, who became the manager of the works; Adolph Miller, Adam Traub, and Mike Shepner, all glassblowers; and William Thompson, blow-smith.

All told, the plant employed fourteen blowers, seven "gatherers," and about thirty other helpers. It was a primitive operation. Proud they may have been of their "corporation," but the fact was that they were still making glass by the open pot furnace and hand blowpipe method, a technique of the trade since first discovered by the ancient Phoenicians.

It was at this point, with a furnace completed by Harry L. Dixon, a glasshouse engineer brought in from Pittsburgh for the purpose, that a major discovery was made.

The Balls discovered that the patents had run out on a type of glass jar similar to those they had been making since 1884 in their first small furnace, a jar used for home canning. Home canning was nothing new to Americans. In 1810 a Frenchman named Nicolas Francois Appert, a man of many parts—chef, brewer, confectioner, vintner, distiller, pickler, and general man-about-foods—put all those parts together and

discovered that food heated and sealed in glass bottles was preserved over long periods of time. His book, *L'Art de conserver les substances animales,* was published in America in 1815. Shortly thereafter, commercial canning here picked up quickly, helped along by Thomas Kensett's first American patent on a tin canister for preserving foods—especially lobster, salmon, and oysters—which he and a partner, Ezra Daggett, started putting up in New York. (The word *canister* was soon abbreviated to *can* and established in the language to connote sterilizing and sealing of food in either glass or tin containers.)

The most famous of these containers was the Mason jar, named for John L. Mason, pioneer inventor in the field. His patent on the screw top for fruit jars, November 30, 1858, was the basic one in the industry. But 21 years later it expired, and the bright young men in the Ball brothers' office realized that the patents on both the "Mason" jar and the zinc cap, so necessary to seal it, had expired. If one company were to make both the jar and zinc cap, reasoned the brothers, a quality control could be set up giving the best possible fit for a jar and cap, thus assuring the canner successful results.

The Buffalo team acted fast. By 1886 they had moved into the Mason jar business and produced "Buffalo" jars with their own name and "Buffalo, N.Y." pressed into the glass lids or in the glass liner of the zinc cap. In 1886 they produced 12,500 fruit jars. Until then the fruit jar business had been monopolized in the United States by the Hero Fruit Jar Company, Philadelphia, Pennsylvania, and the Consolidated Fruit Jar Co., New Brunswick, New Jersey, which controlled the Mason Improved fruit jar and porcelain-lined cap jar patents. These two companies fought their new competition by notifying the trade that the Ball brothers were infringing on their rights, and that they would sue the Buffalo firm and anyone handling the Ball jars to the limits of the law.

The Balls responded with brisk countermeasures. They hired patent lawyers to confirm the fact that the Mason patents had moved into the public domain. They circularized the trade with this information, offering, in addition, to protect any of the buyers and handlers of their jars against any damage or infringement suits.

No suits were brought. The Balls had won. Their wisdom and acumen in combining the production of the Mason jars and the zinc caps were to give them the jump on their competition. But there was more than that to it. They ran the others right off the track with their unrelenting zeal and talent for personal persuasion.

Historian of the glass business Julian Harrison Toulouse caught the true gist of what was happening behind all the legal and marketing maneuvers: "Perhaps the brothers developed salesmanship and marketing while at Buffalo and before they made any glass. As buyers of glass jugs, they became customers of the glass industry. They made these jugs into an article of commerce that they had to get out and sell. . . . Ball Brothers was a whole family of salesmen and could profitably cover a wide area. . . . The sales path was through jobbers, wholesalers, and commission houses, on the way to the retailer and the housewife. They learned the proper sales routes."

Toulouse seems to be shaking his head ever so slightly as he goes on to detail their future triumphs, as if to wonder at the chemistry of success generated by those amazing Ball boys. He could readily see that theirs was an enterprise started at the right time by the right people. What was to turn out to be the right *place*, after the lighting of their first furnace fires in Buffalo, was just ahead in time.

Furnaces, unfortunately, *are* fire. And once again, this implacable enemy, which devastated so many plants throughout the years, before the availability of powerful fire-fighting equipment and preventive fireproofing of buildings, struck Ball.

This time it could have been arson. Their plant was located, in 1886, near railroad yards where there were many saloons frequented by some tough elements. The Balls had incurred the hostility of some hard characters by ordering them out of their factory as they wandered in for a little boozy sightseeing. This last fire completely destroyed the Ball plant. It was little consolation that there was some evidence it had been started by the outsiders who had been given the boot.

No matter. It was done. The Balls were fully insured, and in their indomitable way set to work again to rebuild. It was a cruel setback. All the insurance would not pay for the lost time and interruption of business.

They also suffered a sentimental loss. Their favorite little horse, Fanny, happened to be in the building at the time of the fire and was burned to death. Frank Ball said: "If possible, we regretted her death more than the loss of the factory. She had been so faithful and we had all grown to love her dearly."

Now a new element came to the surface in their business outlook. For some time, the brothers had been hearing that natural gas was being piped from Pennsylvania's abandoned oil wells to Pittsburgh, supplying a new kind of fuel for manufacturing and domestic purposes. A boom had started in the Midwest in communities which suddenly realized there was much to be done with this free fuel coming out of the ground. Citizens formed syndicates and purchased farmlands around their cities. They then advertised for manufacturers who would locate in their midst and who would be given free land, free gas, and enough money to pay for moving the plants.

Some factories were already moving. Land prices were zooming. The boom was on in Ohio and Indiana.

Frank Ball, on the road again, learned from a customer in Cleveland that Findlay, Ohio, was one place he could see a burning gas well and enjoy all of the excitement of a boom town.

Findlay rolled out the red carpet for Frank. So did other places like Bowling Green and Fostoria, Ohio. What he saw made him realize that perhaps the time had come to break up home and family ties in Buffalo. He knew full well that the burning of coal, which was the fuel their furnaces were using at the time, was a very large part of the expense in glassmaking. Perhaps the time had come, once again, to move to "greener pastures," as father Lucius had always been willing to do when greater opportunity was offered.

The air was full of siren songs in that pivotal year of 1886. Many of them were wafted toward the Balls, particularly from Bowling Green, Ohio, where, not strangely, Frank was looking for a sign as to whether he and his glassworks would be welcome. Just at this point, prophetically, Frank received a telegram from someone signed James Boyce, Muncie, Indiana, asking him to come visit there before deciding on a location.

Frank Ball recalled the incident in later years with wry humor.

"I had never heard of Muncie. I looked it up on the map and found that it was a town of about six thousand population about fifty miles northeast of Indianapolis. Having become weary of the monotonous life in Bowling Green and ready for a change, I decided to run down to Muncie and see what they had to offer. I had just been told that the gas pressure in Muncie was very weak, and while they might have enough for cook stoves, it would be folly to think of operating a factory with Indiana gas. I therefore was not very enthusiastic about Indiana, but I had reached a stage when anything was better than staying in Bowling Green."

And so Frank Ball came to Muncie, Indiana.

Part Two

The Pioneer Period

The term "pioneer period" is more aptly used in reference to Ball Brothers than is usual in such company histories, in that the five founding brothers were sons in a family which had emigrated from Canada to Ohio, then lived in New York State under conditions which re-enacted the hardships and simple pleasures of a rural America being carved out of the wilderness in the eighteenth century. The pictorial record of the Ball family enterprise starts with the parents of the five brothers and then wends its way through the family's early years on the shores of a lake in upper New York State to the establishment of their business in the heart of Buffalo. It is a cameo of early American life, a record of thrift, vision, and astonishing dedication to the ideal of success in business which typified America in the latter half of the nineteenth century. It was an era when farm boys could build a great industrial enterprise if only they had the stuff of which Horatio Alger novels were made. And that the five brothers had—they were bound to rise, fortified by pure grit and a kind of frontier energy distilled into business imagination.

Buffalo Metallic Ware Mfty

70, 72 & 74 Washington Street, — Buffalo, N. Y.

Schuyler Aldrich, Proprietor.

W. C. SMITH, Superintendent.

All Articles, Good as Heretofore, with Improvements, and Additions, which make this Firm's line of Tea-Kettles the most peerfct in the World. (See other Side.)

ALL COPPER.   HALF COPPER.

These works spin the bottoms of their kettles into shape from patented pits ; so, that they are substantial throughout, and without soldered seams to open by stove heat and abrasion — the joining being to a central band, above all contact with the stove. The breasts they spin downward and join to the same central band, in such manner that the seams impart strength and beauty to the body of the kettle. The ears and handles are neatly finished in socket joints. The spouts are firmly brazed and strongly set. The finished kettles are highly polished, papered and packed in dozen and half dozen cases, ready for shipment. - - - The wood-clad Cans need no boxing. - - - All other articles are packed as ordered. *Sight Orders, filled in rotation.*

Table Shield, Stove-Pipe Collar and Stove-Foot Rest.

1 Adorns the Table and Shields its Surface against hot dishes.
2 Beautifully relieves the junction between stove-pipe and wall.
3 Ornamentally protects Carpet, Oil-Cloth, Rug or Stove-Board.

ALDRICH'S PRICE LIST, ETC.

All Kettles wanted, and especially in August, September and October, should be ordered shipped on the required day, as long before that day as possible : as orders sometimes exceed the ability to fill promptly. The following Order-Form may be used, writing the number wanted in the blank close after the price of each article. To prompt firms, ordering heavily, suitable discounts are given.

To S. ALDRICH, 74 Wash'ton St., Buffalo, N.Y.

In accordance with the following prices, subject to   ℔ discount and about   1880, please send me by

To

Pit Bottomed Copper and ½ Copper Kettles, per Doz. :

5 in., ½ Cop. $	All Cop. $21.00	Nickel $30.00
6 " "	" 24.00	Plated 33.00
7 " " 18.00	" 27.00	" 36.00
8 " " 19.00	" 30.00	" 39.00
9 " " 20.00	" 33.00	" 42.00

Flat Bottomed Copper and ½ Copper Kettles, per Doz. :

5 in., ½ Cop. $	All Cop. $21.00	Nickel $30.00
6 " "	" 24.00	Plated 33.00
7 " " 18.00	" 27.00	" 36.00
8 " " 19.00	" 30.00	" 39.00
9 " " 20.00	" 33.00	" 42.00

Nickel-Plated Table Shields, per Dozen :
4 inch $2.00,   6 inch $3.00,   7½ inch $3.50,

Nickel-Plated Stove Pipe Collars, per Dozen :
5 inch $1.60,   6 inch $1.70,   7 inch $1.80,

Nickel-Plated Stove Feet Rests, per Dozen :
4 inch $1.70,   4½ inch $2.00,   5 inch $2.40,

Brass Stove Feet Rests, per Dozen :
4 inch $1.50,   4½ inch $1.75,   5 inch $2.10,

Wood-Clad Cans, Each ;
1 Gallon $0.40,   2 Galion $0.60,   3 Gallon $0.70,
5 Gallon $0.90,   10 Gallon $1.50,

S. ALDRICH, 74 Washington St., Buffalo, N.Y.
C. B. JAMES, 97 Jefferson Ave., Detroit, Mich.

A stepping stone in the career of the Ball brothers was provided by a member of Uncle George's church, who conducted the Aldrich company in Buffalo, producing copper tea kettles and wood jacket oil cans. In 1880, Dr. Ball was able to place Edmund as a worker in the Aldrich plant, and soon the owner came up with a proposition—he would sell the wood jacket can part of the business (which he did not mention had been losing money) to young Edmund on a cash value deal, $200, which would make him the proprietor of some hand tinner's tools, a few boxes of tin, and some wood jackets. It was then that Uncle George, the good angel of the brothers, provided the money, and from this modest beginning, the Ball Corporation of the future started its rise.

Edmund Ball called upon his brother, Frank to join him in the new venture. They rented a loft over a paint store on Hanover Street and proudly hung out their first shingle. They learned their new business fast. As the advertisement proclaimed, they made from one- to ten-gallon sizes, and offered them to a wide variety of trades—ship chandlers, lumber camps, and paint dealers among them— which might find them useful.

The Ball boys knew how to handle machinery for sawing up wood from past experience on the farm, and from one former ill-fated venture in the manufacture of wooden kits for packing fish, which had gone up in the smoke of a sudden fire. Now they employed a tinsmith to handle the metal working, and they set out to find the sales which had eluded their predecessor.

They had been obliged at first to pay him a royalty on a method for fastening the jackets to the cans. But soon they invented their own way, soldering the units together with strips of tin. The result was not only a better product but a cheaper one. With this competitive advantage, they were soon supplying all the jacket shipping cans used in Buffalo, and taking orders in Cleveland and Detroit, Indianapolis, and St. Louis. The fledgling business was well on its way.

Lucius Styles Ball (1814-1878) and Maria Polly Bingham Ball (1822-1892) were the parents of the five brothers who founded the family business in Buffalo, N.Y., 1880. They were married in Orwell, Ohio, in 1846. Curiously, both were born in Canada, he in Ascot and she in nearby Stanstead, but did not meet until their families immigrated separately to Greensburg, Trumbull County, Ohio—he in 1834 and she in 1840. There Maria taught school and young Lucius helped his father establish a farm and branched out into a number of business enterprises for himself. Son Frank later wrote of this lady whose inner strength and spiritual qualities shine through the cumbersome photography of the time: "She was a remarkably strong woman physically and mentally with a sweet Christian spirit. Her children and her home were all and all to her. Her chief ambition was to bring up her children with good strong minds and bodies with right ideals and true Christian principles. She urged us to stick together and help each other; it was through her influence and her advice that we five brothers have been associated in business throughout life." Lucius was a striking individual— tall and dashing, his deep-set eyes twinkling with cheerfulness and good nature. He was known by his friends and neighbors respectfully as "Squire Ball." His judgment and fairness were so well considered in the area that in those days before an abundance of lawyers and the availability of law courts, he was frequently called upon to settle disputes. His decisions were generally accepted.

This winsome twosome are five-year-old Frank Clayton Ball (future founder and president of the firm for 63 years and his sister Frances, at only 2 years, taken in the year 1862 by an itinerant photographer passing through Greensburg Four Corners, Ohio. The antique metal picture frame, with matting of fine fabric which set off daguerreotypes of that period, add a touch to the piquant childhood portrait.

Frances (below) grew up to attend Hillsdale College, Hillsdale, Michigan, where she met William J. Mauck, at that time professor of classical languages there. In 1884 they were married. Later Dr. Mauck became president of the college, as did their son Willfred Mauck. A beautiful dormitory at Hillsdale is named in her memory.

Lucina Amelia Ball (older sister of the five Ball boys), was born in Greensburg, Ohio. When the family lived in the Buffalo area, she attended school there, and later taught in Canandaigua, N. Y. She worked for a time as an editor of *The Baptist Union*, a religious paper in New York, and also for Charles Scribner & Sons, publishers. In 1888 she became secretary of Pratt Institute in New York City, the first complete vocational training school in America. In 1891, her educational attainments enabled her to become first secretary in charge of the curriculum at the Drexel Institute in Philadelphia.

Uncle George H. Ball

There were giants in those days, and George H. Ball, uncle of the five boys, was a towering personality among them. He was a Baptist preacher of renown in Buffalo in the 1870s, an educator and publisher. It was his guiding hand—not to mention the loan of $200—which put the Ball brothers in the undertaking which was to reach outer space 100 years after its humble beginnings. Viewing his commanding presence and searching gaze in this photograph, we are not surprised at his vigorous motto: *"Ought it to be done? If it ought to be done, it can be done, and how far can I help in accomplishing the work?"*

Lucius L. Ball   Edmund B. Ball   Frank C. Ball   George A. Ball

These handsome young gentlemen, photographed in 1872 (William C. appears to have missed the sitting), were residents of Canandaigua, N. Y., where the family had moved in 1868, following their years on Grand Isle, in the St. Lawrence River, and in Tonawanda, on the New York bank of the river, a few miles above Buffalo. Now they lived on a 10-acre tract right at the foot of beautiful Lake Canandaigua, one of the Finger Lakes, a sweep of water about 18 miles long, and an average of a mile wide, with the appearance of a Swiss Lake. They skated on the lake in the winter, boated and swam and fished in summer. They trapped muskrats and peddled the skins for 25 to 50 cents apiece, picked strawberries and grapes on neighboring farms, and hired out for chores. They even experimented with raising tobacco and making brooms. Life was rugged but 'most everything was fun.

The Old Academy, Canandaigua, N. Y.

At one time or another, all of the boys and their sister Frances attended Canandaigua Academy. This was a remarkably well-rounded private institution with a large faculty well grounded in considerably more than the fundamentals, a sizable library, and a curriculum which included courses in bookkeeping and business administration. In our terms, it seems to have been something like a junior college. The brothers five quite obviously never stopped learning during their entire lives, but the academy gave them a solid educational groundwork.

The new business had the propitious result of reuniting the five Ball brothers. Lucius and George dropped what they were doing and joined Edmund and Frank in Buffalo. Will sold the Ball homestead in nearby Canandaigua and brought their mother to a new family home in Buffalo. It was in these years that Lucina was working in New York City, and Frances was attending Hillsdale College. Now came the big moment which they had hoped for so long. The firm name was incorporated as Ball Brothers Glass Manufacturing Company in 1886.

The photo below shows the last factory building occupied by Ball in Buffalo, a picture taken when the old narrow structure was serving as a warehouse.

The span of a vehicular viaduct soars over the original spot where "Ball Bros." settled into their offices and plant at 55 Main Street, at the corner of Perry, when the brothers were united in their own business establishment in 1880.

Chapter 4

MOVING AHEAD

Muncie in the late 1880s was caught up in the regional enthusiasm over discoveries of natural gas. On May 7, 1887, headlines in the *Muncie Daily News* proclaimed: "The future of Muncie grows brighter; everybody happy with prosperity; the building boom to begin soon; Gas ... Gas ... Gas ... "

"Nowhere in Indiana can a traveler go but he must be full of gas talk," the newspaper article stated. "And above all other places, Muncie takes the lead as a gas center."

The Muncie gas boom might have started ten years earlier if only a couple of railroad men had been aware of the true significance of a phenomenon they stumbled upon by accident. In the spring of 1876, at a point a few miles north of Muncie, the two were directing the drilling of a test shaft, hoping to find a vein of coal. They were George W. Carter and W. W. Worthington, superintendents of the Fort Wayne and Southern railroad, and they had been at it for over a month without success.

Worthington: "We're wasting our time. We're down six hundred feet and there's no sign of coal."

Clark: "Well, we've gone this far, let's keep boring until Saturday night. Then if we haven't brought up coal we'll abandon the shaft."

At that moment there was a muffled roaring and rumbling down in the shaft. The two men were aware only of some subterranean distur-bance and a strange smell coming from the diggings. It just didn't occur to them to suspect that there was gas down there under the Indiana soil. They shrugged off the rumbling, and since there was no coal, the shaft was sealed off, and the mysterious vapors were locked once again within the earth, without anyone conjecturing what that incident could portend.

The delayed excitement over natural gas began after an 1884 fuel find in Findlay, Ohio, a discovery which led to a burst of prosperity for the town. Hoping to copy Findlay's success at attracting big business through low fuel costs, energetic communities like Muncie began drilling wells throughout Indiana and Ohio.

So Muncie, then a newly chartered city, became the first Hoosier community to make a significant find when fuel was discovered at a depth of only 922 feet in September, 1886. Further drilling ventures proved successful. Thrilled at the seemingly endless opportunites the precious gas could bring the area, residents engaged in dreamy speculations. To the citizens of what was once "Munseytown" (named after the Munsee clan of the Delaware Indians), a small trading center in the days of westward expansion, it suddenly seemed as if they were to be the heirs to all the riches of the industrial age. Muncie had good reason for the excitement. The town had 20 percent of all the producing wells in the entire state. Excitement flared like a flame.

But also prevalent at the time, and potentially harmful, were land speculations. By May 17, 1887, it was estimated that $2 million worth of real estate had changed hands in the preceding six weeks. The Board of Trade was meeting every day because of the volume of work. In a small town where a corset cost only 25¢ and a cord of stove wood 75¢, the newly inflated land prices seemed out of place and possibly dangerous. But Muncie was confident of its eventual success.

On May 11, 1887, the Muncie newspaper ran this article under the headline: "VIEWING THE GAS ... 1,136 Strangers in the City. Men of Wealth and Influence Seeking Locations. Real Estate Men Busy and All is Serene."

"An excursion train started from Cincinnati this morning over the Bee Line gathering passengers from Dayton, Springfield and Greenville and other intermediate points and arrived in Muncie shortly after noon, landing over 1,000 people in our city, eager to see the beautiful uses and effects of natural gas. According to the conductor's count, there were 1,136 people on the train. Mr. Healey, who has personal charge of the excursion, said that he never saw people so anxious to get a place as these excursionists were. All had heard about the wonderful gas fields at Muncie and they were impatient to get there. The first thing each passenger noted was something that we did not notice—the odor of gas. The air of the town is charged with the odor of natural gas which we do not feel, but a stranger detects it immediately. When the crowd of people left their cars and turned on Walnut Street, they passed under the torch on the Bee Line and were almost burned by the intense heat. The crowd has now scattered about the town and are viewing the uses to which the gas is applied. The train will leave for the various places on its return trip at 9 o'clock this evening.

"Muncie is today the best known natural gas town in the world, and there are many visitors daily seeking investments within her limits proving the assertion that boom has come to stay, and those who have purchased property are not in a great hurry to let go as values are sure to go higher in the future. The twiddle twaddle and the gush of the small places around this portion of the state is of no avail."

At the forefront of all this excitement was a remarkably energetic man by the name of James Boyce, the same one who had sent the telegram to Frank C. Ball, inviting him to come see for himself what a wonderful community Muncie was to settle in. Boyce was virtually a one-man Chamber of Commerce.

It was he who built the first electric light plant in the town. That was in 1885, the year the courthouse was built. When the first commercial club was organized in the town, the old Muncie Board of Trade, in 1887, his was the first name on the roster of the incorporators. He had a high sense of initiative.

He was born in Ireland in 1833 of Scottish descent, immigrated to America as a young man, learned the flax mill business in upper New York State, and moved on to Muncie in 1870. Some idea of his entrepreneurial energies may be realized as he not only conducted a successful flax business, he also ran a handle factory, a basket factory, a chair factory, and a rivet mill. Boyce was always first with every new improvement, putting in his home the town's first bathtub and steam heating, even owning Muncie's first lawnmower. Inevitably, he was the leading spirit among Muncie citizens in establishing a fund to induce manufacturers to settle there and take advantage of the new fuel. He himself contributed $12,000 and the fund eventually swelled to $120,000.

Far ahead of his time, Boyce established what we call today an industrial complex on the east bank of the White River, and named it after himself, Boyceton. Said the *Muncie News:*

"Boyceton booms and dwellings are being erected at a rapid rate. Some one or two dozen being now underway. Mr. Boyce has in contemplation the erection of a business block in his new addition. There are no flies in Boyceton."

There seems to be on record only one dissenter to the dreams of Mr. Boyce and his fired-up associations. A certain Mr. J. N. Templer arose during the committee discussions and said, "The discussion is founded on wind backed by wind and will end in wind."

Templer was against this giving of funds to manufacturers. He called it a bribe in which the giver was equally guilty with the receiver and in nine cases out of ten he would "scratch the back of those who desired to locate factories and find 'fraud' written beneath their skin."

But Templer was hooted down and our friend Mr. Boyce prevailed. Well that he did.

Later on, after the dust had settled, the report of the state geologist for 1888 listed no less than six sizable factories using natural gas for fuel:

Muncie Pulp Company; capacity, 20 tons daily; 80 employees.

Muncie Combination Manufacturing Company; capital, $25,000; 25 employees.

Ball Glass Works, fruit jars, green and amber bottles; two furnaces, nine pots; value of daily product, $700; 125 employees; weekly payroll, $1,200; sand from Millington, Illinois; lime from

Fostoria, Ohio; and soda ash from England.

Hemingray Glass Company, bottles; one furnace, fourteen pots; 100 employees; weekly payroll, $800.

C. H. Over, window glass; two furnaces, sixteen pots; weekly capacity, 1,400 boxes of glass monthly; 120 employees.

Muncie Nail Company, steel and iron nails; daily capacity, 500 kegs of nails; 200 employees; monthly payroll, $10,000.

One reader of the *Daily News* expressed in verse his excitement at just being alive during this important time in Muncie's history:

"Tell me not in mournful numbers
That the town is full of gloom,
For the man's a crank who slumbers
In these bursting days of boom."

The fact was, however, that, while the citizens were accomplishing all this, intoxicated by the grandeur of the word *gasopolis*, for all its bustle Muncie yet had quite a way to go to be a metropolis. Despite its big dreams, inspired by visionary ideas of industrialization, Muncie in 1887 was still in many ways a very small town.

An ordinance had just been passed to handle a quite rural problem, prohibiting cows from grazing, not merely on Washington, Main, and other paved streets, but in the river bottom, the open commons near the railway, or anywhere on the several unused acres within the city limits.

When Frank Ball came to town, he found the streets "dusty and very dirty." "There was nothing about the town that particularly appealed to me, but the men were all courteous, kind and businesslike," he wrote. The man who, with his brothers, would someday share real prosperity with Muncie in glass, not gas, seemed more impressed with the good farmland surrounding the area and the quality of the local people than the "God-given boom," as a newspaper called it.

Unperturbed, the citizens of Muncie and the local newspapers continued to indulge in high hopes and hyperbole. The *Daily News* gave this account of Frank Ball's coming to Muncie:

"Upon Mr. Ball's arrival at this place, he at once beheld a city already built, streets already paved, business houses with bustling businessmen already established, with numerous gas wells pouring forth millions of cubic feet of natural gas only waiting for the enterprising manufacturer to come and use it."

While Frank debated the move, the local newspapers reported the possibility in detail. On July 8, 1887, the *Daily News* divulged that he was in town "looking over our gas field." The press then confided that he proposed some arrangement with the manufacturing committee of the town's Board of Trade. On July 27, a more complete story was printed cautioning that the deal depended on assurances that a railroad track would connect the plant with the main line. The next day the big story broke with a headline declaring "They Come." A six-column illustration of the layout of the factories accompanied the article.

The actual deal behind the news involved a contribution of $5,000 from a citizens' fund to help the Ball brothers relocate. A factory site of seven acres was donated and a gas well and private railroad line were promised.

The newspapers reported the new factory would employ 300 persons at "the highest of wages" and would be located on the Galliher farm one mile southeast of the courthouse. The new site would consist of seven buildings occupying three square blocks and covering 40,000 square feet. Most of the structures would be only one

story high for the convenient handling of large amounts of the glass goods.

There were no "bad news" accounts in the papers of the complications which arose to put off the opening of the glass plant. But when the factory was constructed, its railroad track was not built as promised because of a dispute between town and railroad officials. Frank wrote:

"I considered it unsafe to do any more work on the plant before knowing positively that a connection with the railroad would be provided. I therefore locked the doors, nailed up the windows, and left the plant in charge of our cousin."

In this interim, there were only the briefest allusions in the papers to events between the "rail complication" and the final opening of the plant. There were reports that the company had received an order for $20,000 worth of bottles. On January 21, 1888, Frank went to Buffalo "to enlist skilled workmen." However, on February 18, fires for glassmaking started at the factory and on March 1, 1888, the first glassware was produced. (Eventually, the switch line was built.) The *Muncie Times* reported:

"Steam was raised yesterday at the Ball Glass Works and today the first glassware ever manufactured in Muncie was produced. . . . The machinery is all entirely new and was started off in a very satisfactory manner. Today they have been manufacturing coal oil cans and lamp chimneys and Mr. Ball, who has been in the business several years in the East, pronounces them all of superior quality."

Although glass production was newly concentrated in Muncie, the Buffalo metal plant continued operating under William C. Ball, while another small factory, at Bath, New York, produced tin oil cans.

But it would be Muncie that would become home for the Ball brothers now, a small, quiet community of Midwestern folk intoxicated with visions of boom and industrial greatness. The Ball brothers could and would help them, but the transformation might seem to take place overnight, in contrast to the slow, steady pace of rural existence. One Buffalo worker is said to have asked, upon arriving in Muncie, if he were still in the U.S. and if two cents would carry a letter from Muncie to Philadelphia.

Members of the Ball family experienced the same feelings of wonder and even fear at first. A favorite family anecdote involves Mrs. George Ball's coming to Muncie. Before leaving on the trip, she was informed by friends that she was going into Indian territory. During the journey, she witnessed fires dotting the countryside at night as the train moved westward. When she arrived, she enthusiastically told her husband that she had seen "the fires of Indian encampments." He laughed and informed her that what she saw were the flambeaux of the gas wells.

But with his family and former Buffalo employees comfortably settled, Frank would write: *"I was afraid the men would be dissatisfied to leave the life they were accustomed to in the city and come out to Muncie and practically live on a farm.*

"I was agreeably surprised, after they came, to see how contented they were. They seemed glad to get away from the excitement of the city and be where they could breathe the fresh air and, to some extent, begin life over again."

After 1888, when the plant opened, there was no more talk of a boom in Muncie. The essence of what they had observed the year before turned out to be merely get-rich-quick land deals that did little to benefit the community. Citizens now talked of other matters and the newspapers returned to covering the routine events of the day.

Once the hoopla died down, real progress began. In five short years, the town would change dramatically. First would come the paving of all principal streets, then the introduction of a streetcar system that would frighten the horses. With the influence of a new growth industry like the Ball Brothers' Glass Manufacturing Co., Muncie would undergo a more wondrous transformation than even it expected. Even though gas reserves failed in the late 1890s, the city was economically strong and continued to grow.

A BRIEF HISTORY
OF GLASS

"Woven tightly into the pattern of man's history is the glittering thread of the story of glass"—Edmund F. Ball (in his address to the Newcomen Society in 1960).

Prehistoric man used obsidian, a natural glass formed by volcanoes or by lightning striking sandy ground, as a material from which he chipped knives, arrowheads, and other simple tools. However, the story behind man's discovery of glassmaking is still a mystery. A simple version told by the Roman historian Pliny is the most frequently repeated, although it generally is accepted as a fable. Pliny says the art was discovered accidentally by Phoenician merchants who camped one night on a beach in Syria near the mouth of the river Belus. Not having rocks to support their cookware over a fire, they used blocks of natron (soda ash) which they brought from the ship. They slept on the beach after dining and woke in the morning to find that the sand and the natron had fused into a startling new substance—glass.

Another theory has more adherents and more evidence to support it. The earliest method of metal making, which preceded glassmaking in history, produced a rough form of glass as a by-product. It is believed that early metal workers recognized the substance and its potential value and began to develop ways to make glass as a separate product.

The production of glass dates back to the artists of Egypt, Phoenicia, and Rome who were widely celebrated for their work. Of these three nations, Egypt usually is credited with being the first glass manufacturer. The oldest artifacts of man-made glass can be found in the colored glazes used for coating pottery and beads, some dating back as far as 12000 B.C. The earliest pure glass found is a molded amulet colored azure blue and believed to date to 7000 B.C. in Egypt. Other objects made at this time were small bottles and jars used for cosmetics, perfumes, ointments, and even tears from mourners at ancient funerals. Although all of these objects have been found in Egypt, many scholars question if they were manufactured there.

Glassmaking emerged from obscurity to become a definite industry in Egypt during the 12th Dynasty, with the first dated piece, a mosaic glass rod, made during the reign of Amanemhat III (2050-2000 B.C.). Other glass products such as vases and jugs made during this period usually were nontransparent and darkly colored and were made by winding rods of hot, soft glass around a core made of sand or clay. Later it is believed the molds were dipped into pots of molten glass. Until the Christian era, Egypt remained the center of the glass industry. Then it moved to Alexandria, later to Rome and Venice.

The introduction of the blowpipe was the most significant advance in glass production, enabling workers to make articles of priceless beauty and utility impossible to make before its use. When the blowpipe was invented is also a mystery, but most scholars believe it occurred sometime between 300 and 20 B.C. and credit its introduction to the Phoenicians. The standard blowpipe has not changed much since then, still being a hollow iron tube, four to five feet long, with a flare on one end and a mouthpiece on the other. Under the reign of Augustus (27 B.C. to 14 A.D.) the blowpipe caused a revolution in glassmaking and transformed a luxury into a necessity as the art, under Roman rule, spread to the far regions of the empire. Soon glass factories were built in Italy, Gaul, Greece, Brittany, Spain, and Germany.

But with the fall of Rome, after invasions from the north, the great glass centers began to decline. Venetian glass production prospered and for centuries was known as the finest in the world. But elsewhere the skills were lost or forgotten. The glass made during this period was vastly inferior to the products of Egypt and Rome centuries before, and not until the end of the Dark Ages would the art be wholly revived and restored.

It was during the Elizabethan Age that glass production again began to flourish, due largely to the introduction of coal as fuel and the development of lead glass. Large polished plate glass was developed by the late 17th century and was used extensively for windows. Amazing technological progress was achieved a century later by the discovery in 1790 of a method of producing optical glass.

When the English arrived in Jamestown, Virginia, they brought glass to America. The settlers chose a site "in the woods neare a myle from James Towne" and built a "goodly" glasshouse, the first factory in America. Glass was also a part of the colony's first exports authorized by Captain John Smith. While for a number of reasons these earliest producers failed, numerous pioneers in the industry achieved remarkable success as the population grew and demand for products increased in America as well as abroad.

The first phase of substantial progress in the U.S. glass industry came between 1786 and 1815, the result of a drop-off in British imports during the Revolutionary War and the War of 1812. Advances made at this time in the increased number of glasshouses, their financial success, and the extent of their production led to establishing a permanent and healthy American glass industry.

The industry expanded again between 1825 and 1831, keeping in step with the general U.S. economic expansion of the era. By 1860, U.S. glasshouses were believed to be superior to their European counterparts as American ingenuity and planning transformed the small, dimly lit and poorly ventilated shop of 1800 into a factory of superior plant layout and organization.

In spite of the important changes that helped increase the efficiency in glassmaking over the years, the basic process when the Ball brothers entered the business in the 1880s had not really changed much over a period of 2,000 years. In his Memoirs, Frank describes the procedure used at the time this way:

"The process of making glass and producing

jars and bottles in our first open-pot furnace was as follows: The batch consisted of sand, one thousand pounds; soda ash (carbonated soda), four hundred pounds; ground limestone, one hundred eighty pounds. The batch was mixed on the floor by hand with shovels. The pots were filled with batch in the evening, melted during the night, and worked out during the day.

"The workmen were divided into shops—two blowers, one gatherer, one mold boy, one boy to carry from molds to tempering ovens, and one oven laying-up boy. The task of a shop was to work out a pot during the day. In the case of jars, the gatherer gathered the required amount of glass on the end of the blowpipe by revolving it in the molten glass; he then passed the blowpipe to the blower who formed the glass pear-shaped on the end of the blowpipe and then blew an elongated hollow form which was then lowered into the mold and blown to the shape of the mold. The surplus glass was blown over the top and formed what was called a blow-over. The jar was then taken from the mold, placed on a 'buck,' and

carried to the tempering oven. This operation was repeated, the two blowers working alternately with three or four molds."

Starting with the development of F.C. Ball's semiautomatic machine patented in 1898, the intervening years have seen an ancient art progress to a miracle of modern mechanization. Today sophisticated instruments monitor and control virtually every facet of glass furnace operations. Computer-controlled forming machines produce containers at speeds never dreamed of by the glass blowers of a century ago. Invisible coatings strengthen the glass and make possible high-speed filling machines and reduce the cost of food and beverages for consumers. Electronic inspection devices weed out imperfections unseen by the human eye.

Thanks to this modern technology, millions of strong, inexpensive, inert, gleaming glass bottles and jars provide the sanitary, resealable packages and individual showcases for countless products indispensable to our modern life.

Chapter 5

CORPORATE GROWTH

A captivating aspect of the fruit jar business in its early days was that the sales did not suffer from downturns in the economy. On the contrary, in hard times, housewives turned of necessity to the home canning of garden fruits and vegetables. Thus, during the economic hardships caused by the severe panic of 1893, when businesses everywhere were faltering, the Ball Brothers' Glass Manufacturing Company shouldered ahead, prospered, and grew.

The year 1893 was in truth a good one for glass production. Ball company ended the year in a strong position, with a growing reputation for excellence gained just five years after the company moved to Muncie. Now with a work force of 1,000 men and a newly installed Modes continuous tank furnace to replace the eight-open-pots furnace originally used, with additional facilities including a batch mixing room, a department for pressing white liners for zinc lids, metal-stamping works, and nine warehouses, covering altogether 279,800 square feet, Ball became one of the largest green-glass container plants in the "West."

Muncie itself at first appeared largely unaffected by the country's economic disaster, keeping pace instead with the rapid growth of the Ball company, *The Indianapolis News* reported, January 19, 1893:

"Magic Muncie is the name which the inhabitants of the county seat of Delaware County have given to their city. The aptness of the title is appreciated by those who knew the Muncie of five years ago and who know the Muncie of today. There is so little resemblance between the two that old citizens who have recently returned to the place after an absence of a few years have gone about in perplexity, seeking familiar landmarks."

Although the Panic would have been more serious in Muncie had it not been for the factories attracted there by low cost, the Citizens National Bank was at last unable to raise funds on its securities and closed its doors in August. Immediately, however, a group of solid citizens "representing $1 million worth of property" guaranteed the deposits of all Muncie banks and the community was revived. Rents in the city, however, sank to their lowest level in years as landlords cut prices as much as 25 percent. Some factories did close temporarily, but by the fall of 1893, all were back in full swing. Unlike the national effect of the Panic, the overall damage in Muncie was minimal.

In 1893 the first suspicions flickered that the days of an abundant local supply of natural gas were numbered. That winter, pressure was so undependable that oil and coal had to be brought in to supplement the natural gas. Quixotically, there were still some remarkable gas finds occurring here and there, such as one county well capable of yielding 4 million cubic feet of gas per day. But most of the wells surrounding the new find in the Muncie manufacturing area were failing. Some companies were bankrupt, others moved, but most stayed where they were and made do with a switch to coal and oil. Muncie thus prospered in spite of the dwindling fuel supplies, so that between 1886 and 1899 the assessed value of Delaware County property actually increased from $9.6 million to $21.6 million.

There was some unofficial speculation at the time that Ball Brothers would leave Muncie for the more bountiful coal areas "out west." F. C. Ball put some rumors to rest—he acknowledged the dwindling gas supplies, but reassured the citizens: "There may come hard times, but after that a correction will come and then Muncie will be one of the permanent manufacturing centers in the state."

In 1902 the Ball company revealed that it foresightedly had been conducting experiments to devise a better fuel system. In conjunction with the Hemingray Glass Company, also of Muncie, Ball Brothers installed gas converters which could utilize Indiana coal, considered unacceptable for the purpose up until then. Ultimately the company would turn for its non-glass-melting energy needs to electricity and the new diesel engine, exhibited for the first time at the St. Louis World's Fair in 1904.

The decision by Ball Brothers to remain in Muncie, despite the vanishing reserves of the cheap fuel which had brought it there, was continuing evidence that the young firm found the "Magic City" attractive for other reasons. Poster-

ity would reward the Balls for their constancy. Through their own vision and hard work and the conscientious support of the Muncie citizens in their plant, they were soon able to gain outstanding success unrelated to the costs of fuel.

But success is many-faceted. Much of their good fortune in business resulted from continuous studies by the brothers constantly to upgrade company efficiency. To Ed Ball this meant many long hours in the factories, working with his employees at the manual tasks of glassmaking or, as more often in those days, at the exciting challenge of installing a new kind of furnace or machine. To Frank it meant long hours in the office at night, charting new and unproven courses to success. To William, who was frequently on the road selling, it meant going the extra mile and spending lonely nights in distant cities.

An essential ingredient of the Ball formula for success was the fellowship they developed with their employees. To many factory workers, the owners in the front office were simply known as Frank, Ed, Will, and George. Some older employees continued this informality throughout their lives. Others who came to the firm later referred to the top executives as "Mr. F. C.," "Mr. E. B.," "Mr. G. A.," and "Mr. W. C.," still terms of closeness and courtesy.

An amusing story about Ed at work is told by the scapegoat of this tale, an out-of-town businessman calling on the front office for the first time. He was finishing up his business when a "workman" in shirt sleeves (actually it was Ed just returning from some of his frequent labors in the plant) came into the office. One of the brothers told Ed, "Take this man over to the station. And be quick about it. He has to catch a train." When they arrived at the station, the caller tipped the "workman" a half dollar. The next time the businessman returned to Muncie, he was formally introduced to "the" Mr. E. B. Ball at a dinner and immediately, and somewhat shamefacedly, realized he had once tipped one of the famous Ball brothers. There was a good deal of laughter at his expense, but he got his half dollar back.

The Ball brothers' underlying philosophy, however, and a contributing reason to their success can be found in another story, one that Frank was fond of telling even into his old age. It was the oft-told tale of perseverance, known to most schoolboys, about the little train that made it to the top of the hill all the while puffing, "I think I can, I think I can." The story reflects the importance of hard work and a positive outlook, a philosophy that would pay off for the brothers when applied to business large or small, something implanted in them during their early years with their father and Uncle George.

An early display of this philosophy came in 1893 when they purchased the Arbogast Patent No. 269819. The patented invention covered the first departure from the old blowpipe method of glassmaking which, though ingenious, had not been refined by Arbogast to a practical, operative machine. Realizing its potential, the brothers purchased the patent from the United States Glass Company, a manufacturer of fish bowls. They later invented and patented a machine known as "The F. C. Ball, Patent No. 610,515 Sept. 6, 1898 Glass-Blowing Machine." This timely invention completely changed the conventional method of making wide-mouthed jars and bottles from the old-style blowpipe to a Modes continuous tank furnace installed and equipped with Ball-Arbogast pressing and blowing machines.

By 1900 the engineering innovations of the company were beginning to give the brothers worldwide recognition and, of course, this also gave them a competitive edge which was felt in the market. One success always seemed to lead to another and then followed an improved power-operated machine known as the Ball-Bingham semi-automatic, which eliminated the presser, and led in 1905 to the Ball automatic feeder, which fed glass directly to the mold.

The net result of the Balls' initiative in advancing the techniques of glassmaking was, in part, a reduction of costs and an increase in production and profits. In a more fundamental way, however, the brothers made significant and important advances for other glassmakers. Not only were they the first to press and blow a fruit jar, but they were also the first to successfully control compressed air in the formation of a glass container, thereby opening vast possibilities for the glass trade.

In 1908 Owens Bottle & Machine Company developed a machine by which glass is drawn automatically from the melting furnace by suction, formed into finished jars and bottles, and delivered ready to be carried to the lehrs (annealing ovens). The prototype of the Owens machine was correctly judged by the Ball company to be an improvement over the machines they themselves had developed and were using, but the brothers uncharacteristically found it financially impractical to install the better machines. Later the Owens Company improved on its invention, but instead of offering the patent again to the Ball brothers, they granted an exclusive license to manufacture fruit jars to a Ball competitor, the Greenfield Fruit Jar & Bottle Company. Realizing their error and the importance of the modified machine, the brothers started negotiations immediately and purchased the capital stock of the Greenfield Fruit Jar & Bottle Company to obtain the use of the patented machine.

Another important innovation of the young firm, completed in 1910, was to replace the heavy and cumbersome method of shipping jars in wooden cases packed with straw or hay. Ball was thought to be the first manufacturer to pack glass in containers made of corrugated paper. (The brothers also decided to place a dozen jars in a package for the convenience of retail sales.) Happily, it was found that the corrugated paper boxes not only carried the jars more securely than the wood boxes but reduced costs dramatically.

After the new packaging system was adopted, the brothers struck a snag in locating sufficient paper for making the corrugated boxes. Therefore, in 1916, they once more moved with characteristic independence to purchase what was known as the Consumers' Paper Mill, also located in Muncie. By so doing, the brothers exhibited a business attitude that would eventually emerge as the company's far-reaching success formula: If a product or package could be improved by branching out into other industries, then Ball would do it.

It was a foreseeable sequence of this same attitude that led the family into the melting and rolling of zinc, primary ingredient in the lids of fruit jars.

Prior to 1912, zinc used to cap Ball jars was purchased from outside sources. Zinc sheet production was controlled by two companies which were unyielding and arbitrary in their pricing and methods of doing business. The Ball company was obliged to pay cash on the spot, for whatever it bought, at the arbitrary price prevailing at the time of shipping. Predictably, prices went up and

down so that this important cash requirement could not be anticipated accurately.

It just so happened about this time that a zinc rolling mill had been built in connection with a large smelter at La Harpe, Kansas, once again located near a natural gas field. When the gas in that location gave out, the company failed. Ball promptly purchased it and moved the machinery to Muncie. Soon thereafter the company became the first to roll from a slab a coil of strip zinc and use it to form a product. Though it was not realized at the time, zinc would become an important commodity for the firm for years to come, principally in the production of shells for dry cell batteries.

As the young firm expanded through product diversification, it also increased in size in glass-making. By 1904 Ball Brothers was operating glass plants at Converse, Fairmount, Marion, Swayzee, and Loogootee, Indiana; Belleville, Illinois; and Coffeyville, Kansas, all acquired to provide additional capacity for expanding sales. The reasoning behind this series of moves was indicated in later years by George A. Ball to have a logical inevitability:

"It seems that the nature of our business almost from the start, in nearly every line that we are in, has been seasonal and that fact has led by one step and then another to first this new or added line and then another.

"The one-gallon glass body, embossed, tin-jacketed kerosene oil can, which was the first start from tin cans into glass, was a seasonal article, being sold chiefly in the fall and winter months. That fact of being seasonal was one condition that led us to take up the manufacturing of something else in glass and that something else was Mason fruit jars—the glass body fitting in with the kero-

sene oil can and the metal cap fitting in with the metal department where we were still manufacturing tin cans.

"These fruit jars were seasonal, being sold chiefly from April to October, so that it was desirable to have something else in the glass department to manufacture during the other months. That led into the manufacturing of bottles and jars other than domestic jars.

"The fruit jars were originally packed in large, heavy wood cases of six and eight dozen, the protection from breakage being the use of straw and hay. Because the manufacturing and storage finally reached a point where the large quantity of straw and hay necessary to have around the packing plant grew to be quite a fire hazard and, partly to make the handling of the jars easier for the grocer and so that they could be delivered to the housewife in cleaner condition and free of risk of breakage, we changed from packing in the large cases to packing in a wood case of sawed lumber containing one dozen each and protected by means of strawboard partitions. The material for the case was shortly changed to a veneer cut from southern gum and cottonwood lumber.

"When the corrugated paper box was first developed for shipping and thought to be suitable only for express packages, we saw the advantage of that paper case over the wood case and adopted it for the packing of jars even though they were shipped by freight and in carloads and less than carloads. Our consumption of paper finally reached a point where we were at times finding it difficult to get a prompt and regular supply of paper at a steady and satisfactory price. We therefore bought in Muncie a paper mill where the outside and inside sheets used as facing to a corrugated straw paper sheet could be manufactured from imported Swedish pulp and scrap paper. Shipments of these corrugated paper boxes from different box plants in Chicago, Fort Wayne, Cincinnati and other places was a bulky matter, with high freight charges, and the supply was very irregular and we found it advisable therefore to put in our own plant for manufacturing the corrugated paper boxes, buying from mills in the north two qualities of paper required, the Jute outside and inside lining or facing and the straw sheets for corrugating.

"Next, we were finding it difficult to get a steady supply of the straw paper for the corrugated sheet and at a satisfactory price, so we bought a plant at Noblesville for the purpose of manufacturing our own corrugated sheet out of wheat or oat straw. The production of the straw paper mill at Noblesville and the Jute liner mill at Muncie was more than was required for our own consumption for the packing of our glass products. We therefore began selling the surplus to other manufacturers of corrugated paper boxes.

"We first bought the zinc metal for the manufacturing of the screw caps for the fruit jars in large sheets 36" x 84". At times we found an unreliable and unsteady supply of that material so that we developed a new process for rolling zinc, making it into long ribbons or strips and putting it into coils. As this process of rolling was not used by the then existing zinc mills, we built and equipped our own mills for rolling these zinc strips. The output of that mill was more than was required for making the caps of our fruit jars and we therefore began to manufacture other articles made of zinc out of those strips so that we could occupy the capacity of the zinc mill.

"We found that the supply of the jar rings or rubbers, one of which was furnished with every *jar, was not always satisfactory. The right quality of rubber being exceedingly important in order to ensure the safe keeping of the product in the jar, we began making our own rubber rings so that we could control that quality. The making of these rubber rings also was seasonal to quite an extent and we therefore, to give a more steady operation in the rubber plant, began manufacturing other mechanical rubber goods."*

Expansion and progress, however, came not without anguish and turmoil. Fires continued to plague the adventuresome young business. On October 28, 1891, "the most disastrous fire that ever visited Muncie" hit the Ball stamping department, causing approximately $80,000 damage. When the plant was rebuilt, it was fireproof, constructed entirely of brick and steel with cement floors.

Still another fire struck, this one in March 1898, destroying Factory No. 2 after fiery molten glass escaped from a tank. Because the plant contained new press and blow machines not yet covered by insurance, the Ball company suffered a heavy loss. Just six weeks later, a new warehouse containing millions of fruit jars was completely destroyed in yet another setback, causing between $250,000 and $300,000 damage.

According to a newspaper account, the ruins from the fire were described as a two-block-long mountain of green glass 50 feet high. Years later a younger member of the firm recounted what G. A. Ball had to say about this "most critical blow":

"As I recall, he indicated that it was necessary for them to go to Buffalo and obtain the necessary financing to keep the business afloat from the Marine Midland Bank. It was on the basis of their previous performance as good borrowers

with the bank that the officers lent them sufficient money to tide them through the following season.

"G. A. indicated that the following season was a good fruit jar year and they were fortunate in making enough money to pay off the loan in the season following the rebuilding of the facility.

"In this era Ball Brothers had been buying up the stock of the original investors in the company, and it is for this reason among others, I am sure, that they were short of cash to make this rebuilding program out of the company resources."

Other unexpected problems developed, including the now-familiar unrest among workers, then a new phenomenon, as they feared for their jobs when new labor-saving equipment was installed. F. C. Ball reassured them, as in the instance when his glassblowing machine was installed. "It simply is a revolution," he said, "like that caused by the sewing machine, the spinning jenny, the self-binding harvester and other inventions that have failed to injure the workmen, while increasing the demand." He said the only change at Ball Brothers would be that the plant's skilled workmen would now be machine operators instead of blowers.

There were some workers, yes, who lost their jobs to these advances in technology, but for the most part they were unreliable gathering boys who often caused plant shutdowns when they would refuse to work when the circus was in town, just two blocks north of the plant. The more seasoned employees, however, made the transition successfully and found that it improved their own chances for advancement.

Fruit jar sales figures reflect the company's growth during these early days. Despite the severe national economic depression in 1893 to 1895, Ball Brothers produced 22 million fruit jars in one year, beginning September, 1894. The following year's total was 31 million. In 1897 it reached 37 million. By 1905, with 1,200 workers and six continuous tank furnaces equipped with Ball patent blowing machines, the sales figures soared to 60 million jars.

A story in *The Indianapolis Journal* of 1910 gives the output of Ball as an astonishing 90 million jars, "one for every man, woman and child in the United States, with some left over."

The headlines proclaimed that half the world's fruit jars were made in Muncie, with Ball dominating the fruit jar business of the world. The story reported how Ball's central factories and warehouses now covered 35 acres, up from the seven acres of less than 25 years before. Fifty additional acres were used then for other facilities in Muncie and Greenfield, Indiana, and Coffeyville, Kansas. Stored outside at the Muncie plant were "glass jars by the acre, glistening in the sun."

A family anecdote recalls the sea of glass at Muncie, a reflection on the company's flair for success in such a relatively brief period. The story goes that a flight of wild ducks migrating across Indiana on their long trip south for the winter attempted to land on the field of bottles and jars one day, mistaking the glistening mass for a body of water.

Close to the turn of the twentieth century, the Ball brothers seemed to be at home permanently in Muncie. This became evident when the stockholders of the Ball Brothers Glass Manufacturing Company of the State of New York met on March 24, 1899, to sell the company to a newly organized firm, Ball Brothers Glass Manufacturing Company of Muncie, Indiana. All 2,000 shares of stock in the new firm were retained by the five Ball brothers.

At home in Muncie with production up, quality assured, and costs reduced, the Ball brothers began to attract favorable press comments. *The Muncie Times* reported in August 1906:

"To make an article so good that its quality cannot be improved and at the same time at such a low price that other manufacturers do not find it alluring to enter the field seems to have occurred to but a few and one of the few is the Ball Bros. Glass Manufacturing Company whose products are used in every civilized country of the globe."

Although by the early part of the twentieth century the Ball brothers had acquired a far-flung reputation for the quality and quantity of the glass containers they produced, nowhere was their success better received than in their old stamping grounds in Canandaigua and Buffalo, New York. A story in *The Canandaigua Times* by a native, C. A. Briggs, tells of his visit to the Ball brothers in Muncie. He regales his readers with the news that their old friends, who once started out so bravely on a shoestring, now were *"keeping young by controlling a great business, helping the city to grow, . . . diversifying their work by taking over a steamship line on the Great Lakes or perhaps buying up a factory at a figure that would stagger an ordinary speculator."*

The reference made to a steamship line on the Great Lakes was a pardonable mistake. The brothers did not actually operate a steamship line, but they had solid financial interests in the Globe Steamship Company. Their prominence was dramatized on the lakes when in December, 1905, a new $400,000 steamer of 10,000 tons, 550 feet in length, was christened *The Frank C. Ball.*

But our visitor from Canandaigua went on to reassure his readers back home that the Ball brothers had not changed. *"These men have succeed-*

ed in a wonderful way," he wrote, "but their success has in no way detracted from their personal charm ... They are gentlemen of the old school, a type that is seldom seen in business today ... They will always be the 'Ball brothers' who won friends in their Canandaigua home."

Part Three

At Home in Muncie

When the Ball brothers decided on the move from Buffalo to Muncie, they were already recognized in industry. The local newspaper proclaimed the news as more than just that—an event: "They Come." Thus began a period in the community, self-named The Magic City, which was to eventually make the names of *Ball* and *Muncie* virtually interchangeable. Here the brothers built their homes, elegant works of art set in a veritable Eden. The life of the Ball families in their homes on Minnetrista Boulevard was a charming period piece filled with warm relationships and unforgettable tableaus of the times that found their way into the literary works of a famous Muncie author, Emily Kimbrough. South of town, the great glass plant was ever growing, until the Ball brothers would see their product, the commonplace fruit jar, find its way into every part of the civilized world. During those early years of the century, Muncie found its character without losing its privacy. Yet that too moved into another phase as *Middletown* was published in 1929 and Muncie became the prototype of the America of the times.

The name of James Boyce looms large in the Ball saga, since it was he who first sent a telegram inviting the brothers to visit Muncie, Indiana, and investigate the industrial potential of the newly discovered natural gas wells. Boyce was an unusual man, born in Ireland in 1833 of Scotch descent, who immigrated to America as a young man, learned the flax mill business in upper New York State, and moved on to Muncie in 1870. Some idea of his entrepreneurial energies may be realized as he not only conducted a successful flax business there, he also ran a handle factory, a basket factory, a chair factory, a rivet mill and laid out his own "industrial complex" of Boyce-ton. He was a great lover of trees and natural beauty. His residence, shown above, had the town's first bathtub. The discovery of natural gas turned Muncie—if only temporarily—into a boom town. At that point, the vigorous citizens banded together to bring new industries to their area. Boyce himself contributed $12,000 and the fund eventually swelled to $120,000. When Boyce's telegram reached Frank Ball, offering him $5,000 and free land and gas if he would establish the company in Muncie, Frank remarked, "I had never heard of Muncie. Ready for a change, I decided to run down and see what they had to offer." The rest is history.

Muncie, as Frank first saw it, had come a long way from the days of its beginnings as a tenting ground for a clan of Indians called Munsees. When Frank arrived, the new courthouse was being erected. At right is a woodcutting of the local bank as it appeared during the city's infancy. But it was not so much the latest construction that impressed him. He enjoyed the beauty of the local farmlands and he found the citizens to be "courteous, kind and businesslike." The local newspaper recorded his arrival and impressions in detail, but chose to emphasize the more modern aspects of this "gasopolis." It was reported: "Upon Mr. Ball's arrival at this place, he at once beheld a city already built, streets already paved (and) business houses bustling with businessmen already established." In fact, Frank found it an attractive place "to begin life over again."

This woodcut showing the elegant Kirby House, below, illustrates that the community was hardly the "wild west" that Easterners supposed. It was here at the Kirby House that Frank C. Ball stayed on his first visit to Muncie, welcomed by congenial landlord Heinsohn who helped him decide that Muncie would be a compatible place to build a glass plant. Edmund B. Ball made this hotel his home as a bachelor for almost 16 years before his marriage.

The sketches on this page appeared first in *Frank Leslie's Illustrated Newspaper* dated May 4, 1889 (which is excuse enough, perhaps, for a slight indistinctness). They mean much to our story because they were culled from an article on Muncie: "Natural Advantages and Natural Gas Are Calling the Citizens of the World to This Favored City."

"What new wonders has Nature's storehouse given to enrich now fortunate, to be mighty, Muncie, the manufacturing city of the West?

"It is the flame that shoots from Nature's boundless reservoirs lighting up Muncie's future with untold benefits, bringing costless fuel, and setting the wheels of a hundred factories in tuneful motion.

"Its origin is shrouded in mystery. No man knows how it is made or from whence it comes. A well is sunk from 900 to 1,000 feet, and the Trenton limestone is struck; penetrating this rock a few feet, the gas comes gushing up with an almost irresistible force, and when lighted throws a terrific flame high in the air. Sometimes the volume of gas is so great that it is difficult to control, although this is now reduced to a science.

"Thirty-five mighty natural-gas wells, within a radius of two miles of Muncie's centre, pour forth 100,000,000 cubic feet of gas per day, only about one-third of which is now consumed. The amount of saving in fuel and light to Muncie's residents and manufactories now aggregates the enormous sum of $500,000 annually.

"The supply is inexhaustible. Centre Township, six miles square, in the centre of which is located the City of Muncie, will supply gas in abundance for a thousand wells, or two billion cubic feet per day, enough for manufacturing city of one million inhabitants. Taking these facts into consideration, who can anticipate Muncie's magnificent future? Conservative people estimate its population at 50,000 within five years; others at 100,000.

"Among the prominent and largest manufactories located here during the last eighteen months is Ball Bros. & Co., fruit jars, etc., 150 men and saving $10,000 yearly in fuel. It is estimated that the fifty-four factories located here save annually $250,000 in fuel."

The sketchy illustrations show a display of natural gas, the Muncie courthoue (right) and (below) the offices of the Muncie National Gas Land Improvement Co. of which is written:

"The Company owns and controls nearly 2,000 acres of land adjoining the city, and the main streets run through it. The belt-line railroad also passes through the property and streetcars will soon be running. Broad and beautiful avenues are being made, and the land has been laid out into house and business lots, factory sites, and large sales are being made daily."

Edmund B. Ball

Lucius L. Ball

William C. Ball

Frank C. Ball

George A. Ball

The Ball brothers, viewed at the time they arrived in Muncie, successful manufacturers from Buffalo. Edmund, the balance wheel for his brothers' energies and enthusiasms. William, the great salesman, traveler, and raconteur. Lucius, the oldest, who headed the family after the father's death, then became a physician at the age of 40, fulfilling a lifelong ambition after the younger brothers became self-sufficient. George, to whom the others turned for understanding, patience, guidance and advice— whether the problem would be personal, civic, or business. Frank, the born leader, strong character, shrewd businessman, who guided the company until his death in 1944.

In the above photo taken in 1887, before the plant was completed,
the office is in the foreground at the right. On March 1, 1888,
operations commenced in Muncie. The first glass made that day was for
oil containers and lamp chimneys.

The coming of the Balls from Buffalo to Muncie was hailed with delight by the *Muncie News* (below). Headlining the news was the statement that Ball was one of the largest glass manufacturers in the world, a tribute to the business started on a shoestring only seven years before in Buffalo. Production in 1886 had been 1,785,600 jars and 2,520,000 for the following year before the Buffalo factory was destroyed by fire. Sizable enough.

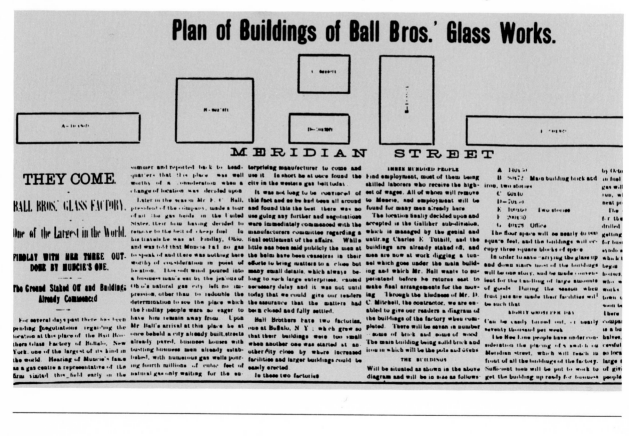

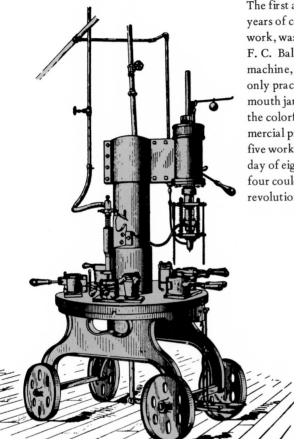

The first automatic glass jar machine, the result of years of costly experimental and developmental work, was completed in 1898, and registered as the F. C. Ball Patent No. 610515. This "miracle" machine, generally recognized by the industry as the only practical machine for manufacturing wide-mouth jars and bottles, signaled the end forever of the colorful and skilled glassblower in the commercial process. Under the hand process, a team of five workers produced about ten gross of jars in a day of eight hours. With the new machine, a team of four could produce 25 gross of jars. The industrial revolution had come to an ancient craft.

This artisan, blowing a gob of molten glass into a mold to form a bottle is performing a task as old as civilization. Although the precise origins of glassmaking are lost in the mists of antiquity, it first surfaced on a large scale in the Egyptian culture, was refined to a superb art under the Romans and, as European civilization grew, it eventually became a basic industry. The blowpipe seen in this illustration was probably invented by the Phoenicians a few centuries B. C. It changed little over a period of two thousand years and was used up to the time machines such as the F. C. Ball machine revolutionized the industry around the turn of the Twentieth Century.

Presented to Mrs. Edmund B. Ball by Cyrus E. Dallin Oct. 6 1928.

The five Ball brothers were to heed, in more ways than one, their mother's wish for them to be together throughout their lives. United, their success in business was the envy of competitors. And when it was time to establish permanent residence in their new community of Muncie after the move from Buffalo, they built their home lives around each other. Finding the environs of the city too confining—perhaps because of memories of their previous country homes in New York State along the banks of the Niagara River and on the shores of one of the Finger Lakes, Canandaigua—they agreed upon an enticing site just outside town on the bank of the meandering White River. It was rich in Indian lore. In fact, a boulder connecting the place with Indian history stands today on the spot with the inscription:

"This boulder marks the traditional site of Wah-Pe-Kah-Me-Kunk, the White River town of the Munsee Clan of the Delaware Indians. It is also a memorial to Joshua, a Christian Indian, who was burned by the tribe, March 1809, a martyr to his faith. In this vicinity the Indians cultivated their maize and corn, and later it was known as the 'May Ground,' where the villagers of Munsee town gathered for their springtime frolics."

The place where all the Balls were to live so happily was to be called Minnetrista. In 1929, the bronze sculpture shown here, *Appeal to the Great Spirit*, a replica of the Cyrus E. Dallin original in the Boston Museum, was dedicated as a memorial to Edmund Burke Ball by his widow and children. It stands at the triangle between Walnut Street and Granville Avenue, where Minnetrista begins.

The fanciful name of "Minnetrista" was first given by the Ball brothers to the site they chose (sometime prior to June 1894) for their stately row of family homes. The name was coined by the brothers' two sisters, Lucina and Frances. To the Indian word *minne* ("water") was added *tryst*, meaning an agreed meeting place. The painting, left, by J. Ottis Adams (brother-in-law of Frank C. Ball) depicts the picturesque site three years before Minnetrista came into being.

Frank C. Ball began building this residence on Minnetrista in 1894. G. A. followed with his home, in a woodland setting, which is still maintained by Miss Elisabeth Ball. Its wildflower garden is widely known for its springtime beauty. Dr. Lucius L. Ball next acquired a farmhouse for his residence, moving it and turning it to face the boulevard, as it does today. William C. Ball then built the Georgian-style home with Palladian entrance. Edmund B. Ball built his residence a few years after his marriage in 1903.

EDMUND BURKE BALL

MISS BERTHA CROSLEY

On October 7, 1903, Edmund Burke Ball and Miss Bertha Crosley were united in marriage at The Central Universalist Church of Indianapolis. The bride's father, the Reverend Marion Crosley, performed the ceremony for his only daughter, and was assisted by none other than Dr. George Harvey Ball, the groom's uncle and president of Keuka College at the time. G. A. Ball was the groomsman, and the ushers were the brothers of the groom. The bride, a '98 graduate of Vassar, had spent much of her life abroad but met her future husband while visiting relatives in Muncie. The newspaper account mentioned that Mr. Ball owned an electric car line in Chicago, a railway in Michigan, and (with his brothers) the largest fruit-jar factory in the world. Then, one encompassing sentence: "For 48 years his heart has been adamant against the arrows of Dan Cupid, and while 30 years ago he was virtually without a penny, today he is believed to be the richest man in the State."

Edmund B. Ball built this imposing residence on Minnetrista Boulevard for his bride between 1905 and 1907, in a beautiful setting high above the White River. The value of the building and grounds was appraised at $100,000, a solid sum for those days. Its architect was the distinguished Marshall S. Mugarin, who incorporated into the design elements of English Tudor. The building is now known as the Ball State University Minnetrista Center for Nontraditional Studies.

To the near right is a photograph of Miss Lucy Ball, daughter of Frank C. Ball, which appeared in the newspapers upon the announcement of her engagement to Alvin Owsley, "formerly national commander of the American Legion, who lives in Texas." According to the newspapers, the news of the betrothal, "created widespread interest because of the prominence of the two persons." At far right is another young inhabitant of the Minnetrista colony, Miss Elisabeth Ball, daughter of George A. Ball. With this picture was the caption in *The Indianapolis Times*, January 23, 1937: "Quiet philanthropies are her diversion."

 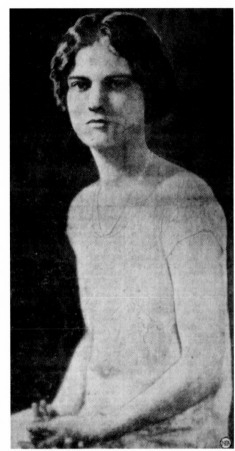

A family portrait finds Mr. and Mrs. William C. Ball quietly ensconced in their living room, characterized by classical touches such as were favored by all of the Ball Brothers—here a decorative fireplace with sculptured mantel, flanked by formal columns. Note also the ornate frames of the mirror and painting over the fireplace, works of art in themselves.

Family life at Minnetrista never was without the fun derived from pets and horses. Unconcernedly cropping the grass is one of the ponies kept to pull the carts taking the children to town or Sunday school. From left to right in the photo: Far left, Mrs. Frank C. Ball (Bessie); Mrs. William C. Ball (Emma); unknown; William C. Ball; Dr. Lucius L. Ball; Frank C. Ball; Helen Ball Robinson.

The daughters of the five Ball Brothers, (known affectionately as "the Ball cousins"), gather on the steps of Frank C. Ball's home. A coat thrown casually over the railing and the relaxed expressions on the faces of these Muncie socialites bespeak the non-posed quality of this picture and these young women. The "cousins," left to right: Helen, Rosemary, Janice (center and front), Elisabeth, Lucy, Margaret and Adelia Ball.

Right, G. A. Ball has taken to one of his favorite pastimes of driving a sleigh pulled by a smart trotter. The view suggests some of the outdoor activities which living outside the actual city of Muncie afforded. G. A., who appreciated fine horseflesh, is said to have made the initial train trip from Buffalo to Muncie in the freight car provided for his horse, to insure a safe arrival.

Their homes side by side along the shady boulevard, the Balls remained a close grouping of related neighbors. Here in front of F. C.'s home is an informal gathering of Ball cousins, aunts and uncles. They are, left to right: "Cousin" Ella (seated); Mrs. Arthur W. Brady (Caroline); Mrs. Lucius L. Ball (Sarah); "Cousin" Julia Ball; Dr. Lucius L. Ball; Arthur W. Brady; Mrs. Edmund B. Ball (Bertha); William C. Ball; Edmund B. Ball; Mrs. Frank C. Ball (Bessie); Mrs. William C. Ball (Emma); Frank C. Ball; and E. Arthur Ball (son of F. C.).

In 1898, after centuries of the same craft processes, Ball led the entire industry in a giant step forward with the introduction of the F.C. Ball machines shown above in Number One Glass Plant Blowing Room, circa 1900. In 1900 and then 1906 new patents were issued, after work by F.C. Ball and A.L. Bingham further refined the original machine, and the Ball-Bingham machine revolutionized the industry. Completely automatic, it did anything except bring glass to the machine and carry the finished bottle away.

Another type of machine concurrently invented by Michael J. Owens, was also called a marvel of glass making (below). Ball at first did not consider using it except on option, until Owens ultimately licensed the patent to the Greenfield Fruit Jar and Bottle Company, acquired by Ball in 1909 and with it the exclusive license to make fruit jars on the Owens machines.

"BALL MASON" FRUIT JARS.

Ball

MASON

There is a difference in the Mason Fruit Jars made by various manufacturers. "BALL MASON" Fruit Jars are the best because they are made of the best material, by skilled workmen. The Jars are carefully selected, provided with perfect Caps and good Rubbers. They are thoroughly reliable and will keep fruit for years.

Why take chances of your fruit spoiling by using inferior or untried brands? The "BALL MASON" has been used over 20 years, is known to be reliable, and can be obtained at a reasonable price. More of them are used than all other brands combined. This fact is evidence of their superiority.

Demand "BALL MASON" Jars.

BALL BROS. GLASS MFG. CO., - MUNCIE, IND.
Largest Fruit Jar Mfrs. In the World.

This 1904 advertisement (above) speaks for itself: "Ball Mason fruit jars are the best." The company designated itself as the "Largest Fruit Jar Manufacturer in the World." The message appeared in the August issue of the *Woman's Home Companion* in that year.

Ball

The trademark granted Ball in 1907 appeared as above, causing some wags to comment that the company did not know how to spell its own name and was using three "ls" instead of the correct two. This Ball script, one of the best known trademarks in the world, inspired much conjecture as to its source. A comparison with the signatures of all of the Ball brothers, as shown at right, suggests a relationship to them all.

Ball's 1912 sales brochure offered four sizes in Mason jars, and a variety of jelly glasses. The oil jug, evolved from the original product in the Buffalo plant, was still manufactured in gallon and half-gallon sizes. In the same brochure, it is noted that the company had factories in Terre Haute and Greenfield, Indiana, and plants in Coffeyville and La Harpe, Kansas, at this time.

In 1903, this is how Ball stored its jars, in a veritable sea of glistening glass. This scene is in the fields south of the Muncie plant as the jars were stored between layers of straw until shipment. A family anecdote tells how a flight of wild ducks, migrating across Indiana on their long trip south for the winter, attempted to land on the field of bottles and jars one brilliant day, mistaking the shimmering reflection for a body of water.

In this photo taken in January, 1903, company president Frank C. Ball reads papers at his desk at the Muncie plant. Frank is remembered as "the ruler of the clan, always stern and serious," but over his solid desk, with its many cubbyholes and drawers is a tell-tale tender photograph of his children. (F. C.'s desk is still kept clean and polished in the company's headquarters, a lingering tribute to the man who led Ball Brothers as its helmsman for 63 years.)

Chapter 6

PATENTS
AND ACQUISITIONS

When G. A. Ball explained in the preceding chapter why the company made certain thrusts into the "outer space" surrounding the glass business—into seemingly unrelated prospecting in the manufacture of zinc, rubber, and paper—his logic was almost prophetic.

The time was to come many years later when the formula repeated itself.

In 1956 Ball made an exploratory investment in a small company with an electronic device that was designed for weighing vehicles and might have application for weighing the raw materials used in glass manufacture. As it happened, the device never proved effective, but it did lead the way, step by step, to Ball's entry into aerospace technology, photographing the mountains of Mars and developing the star tracker system for the space shuttle's flight. G. A., had he lived to experience and savor the ingenuity of Ball in outer space, might well have thought it all very logical.

In 1978, mindful of the multiproduct company Ball had become, Chairman and Chief Executive Officer John W. Fisher described the corporation as "a packaging company with a high-technology base."

Before that could come about, however, Ball had to carefully maneuver through two critical phases in its growth. The first was expansion of the company within the glass industry itself. The second was the far-reaching effect of the final judgment in the case of United States vs. Hartford Empire, et al. in the 1940s, which, in effect, prevented Ball from any further acquisitions within the glass industry. The two sequences are a fascinating study in industrial growth. A company moves imaginatively and aggressively to secure a strategic position in its chosen field. It then runs head on into legal restrictions which are applied almost as a penalty for its success. The company loses an important court battle, but not its determination and momentum toward continued success. It survives the crucial test of endurance the court case caused, emerging years later in a stronger position than ever before through expansion into many other seemingly unrelated fields.

While Ball owes much of its second-phase success to what John Fisher calls "high technology," an early observer could also have noted the same of its earlier growth. From 1898 to 1908 Ball and Ball-Bingham machines were America's most sophisticated glass container machines. Although it may be stretching a point to view the early Ball inventions as "high technology," they should be seen in relation to the relatively primitive state of the industry at that time.

In its hundred-year history, the Ball company

was to progress through four stages in the making of glass—from the primitive to the most sophisticated. We have noted how the industry followed ancient practices in the making of bottles up to about 1890, without making much progress in the creation of special glassmaking machinery. What appeared to keep glass manufacture behind other businesses during the Industrial Revolution was the perplexing and infuriating nature of glass itself, i.e., its peculiar properties in the molten stage, the difficulty in handling it, and a lack of scientific understanding of the chemistry and physics involved. The first significant advance in glassmaking, one which marked the second period of the history of the art, was the development between 1890 and 1894 of the new furnaces, pots, and tanks for molten glass. The third period saw the first radical departure from ancient practices of glassmaking with the Arbogast patent, purchased by Ball Brothers from the United States Glass Company. After the years of costly experiments and developmental work, the Ball company produced what was generally recognized as the only practical machine for manufacturing wide-mouth jars and bottles, the F. C. Ball-patented machine. While in daily operation, the machine's defects were corrected by inventions of Ed Ball and his first cousin, A. L. Bingham, leading to the development of the Ball-Bingham Semi-Automatic Glass Forming Machine, another giant step for the industry at large.

The fourth and most important stage in the history of glassmaking was the introduction of the first fully automatic machine, which was achieved by combining the Ball-Bingham Semi-Automatic Machine with a unique process, also developed by Ball, of feeding the molten glass to the mold automatically. This advance was accompanied about the same time by the invention of another form of automatic glass machine by Michael J. Owens.

Because of the "breakthrough" nature of these Ball initiatives, the brothers' contribution to the progress of glass production was substantial. It was pioneering achievements such as these that kept Ball the leader of its market, ensuring that the company could match the world's best in the manufacture of a superior and needed household product.

There were reminders near at hand, however, that industry competition was not about to slack off, even in the face of Ball inventiveness. Right across Macedonia Avenue from the Ball Brothers' plant was the Hemingray Glass Company, which had moved to Muncie in 1888, also lured by natural gas, and which set about making glass insulators and fruit jars. The Port Glass Company came to Muncie in 1891 and also began making fruit jars, although it moved out to Belleville, Illinois, in 1902 when Muncie's gas failed. The Nelson Glass Company was a strong Muncie competitor from 1893 to 1896. The Muncie Glass Company (later in 1900 the Charles Boldt Glass Company) made beer and beverage bottles at that time and the Boldt Mason Jar. It was based in Muncie from 1892 to 1912, when the company moved its headquarters to Cincinnati.

There were other competitors in the area just before the turn of the century. Among them were the Upland Flint Bottle Company (owned by A. M. Foster), the Safe Glass Company, and the Upland Cooperative Glass Company, all of Upland, Indiana. Others included the Redkey Glass Company of Redkey, Indiana; Pennsylvania Glass Company of Anderson, Indiana; Illinois Glass Company of Alton, Illinois; and Root

Glass Company of Terre Haute, Indiana. An employee of that last firm went on to design the famous Coca-Cola bottle. Ball Brothers' staunchest competition came from two firms outside the celebrated gas belt; the Kerr Glass Manufacturing Company, with plants located in Altoona, Kansas, and Sand Springs, Oklahoma; and the Atlas Glass Company, later known as Hazel-Atlas. The latter company was located in Washington, Pennsylvania, and Wheeling, West Virginia.

Beyond these companies, on the national scene, were a host of other glass factories striving and straining to get ahead, as the country's population grew and the demand for glassware jumped with no apparent limit.

It would appear that the Ball brothers, in touch with all of the markets of the United States through their vigorous sales contacts (as someone had remarked in the Buffalo days, these five brothers were unbeatable as a sales organization), perceived that another road to success was a variation of the old "if-you-can't-beat-'em-join-'em" rule of thumb. The Ball variant of this was acquisition, either for the purpose of expansion or to acquire, in certain cases, equipment or patents otherwise unavailable to the Muncie plant. A review of these far-flung acquisitions is revealing, not only for the vigorous thrust of the brothers in Indiana, but for the impressive funds their business was now providing for such investment, when a dollar was a dollar.

A careful examination of the purchases and when they occurred, in conjunction with operations built independently by Ball, best illustrates the company's first-phase expansion. The most important of these acquisitions, for a reported $600,000, was that of the Greenfield Fruit Jar and Bottle Company for the purposes of acquiring the use of the Owens automatic machine, and patent rights for the manufacture of fruit jars, improved in part by Ball technology. This glass-making wonder had 10,000 parts and weighed 100,000 pounds. With this new machine and the family business sense, the Ball brothers would clearly emerge as the dominant producer of fruit jars.

In this first-phase expansion, we find acquisitions as follows:

1898 *Ft. Wayne Glass Works, Upland, Indiana.*
1901 *Windfall Glass Co., Windfall, Indiana.*
1904 *Marion Fruit Jar & Bottle Co., Marion, Indiana.*
      Fairmount, Indiana.
      Converse, Indiana.
      Coffeyville, Kansas.
1904 *Port Glass Works, Belleville, Illinois.*
1904 *Loogootee Glass Works, Loogootee, Indiana.*
1904 *Swayzee Glass Co., Swayzee, Indiana.*
1904 *Upland Glass Co., Upland, Indiana.*
1909 *Root Glass Co., Terre Haute, Indiana.*
1909 *Greenfield Fruit Jar & Bottle Co., Greenfield, Indiana.*
1909 *Mason Fruit Jar & Bottle Co., Coffeyville, Kansas.*
1913 *Wichita Falls, Texas, new plant.*
1925 *Schram Glass Manufacturing Co., St. Louis, Missouri (office only there— plants at Hillsboro, Illinois; Huntington, West Virginia; and Sapulpa, Oklahoma).*
1929 *Pine Glass Co., Okmulgee, Oklahoma.*
1936 *Three Rivers Glass Co., Three Rivers, Texas.*

Prospects were enhanced by the largest of the

acquisitions listed here, the purchase of the three Schram plants for a total of 2.8 million. While Schram was a major competitor of Ball Brothers, the move was inspired mostly by the acquisition of some 45 patents. Although it proved a futile effort, Ball Brothers wanted the patents for protection against a potential outside patent monopoly.

Meanwhile, Ball was showing its packaging flair in materials other than glass—again to its own advantage. In 1894 the company made its first wooden boxes, of Michigan pine, for shipping jars in dozen lots to grocers. Strawboard liner-dividers protected the jars. This was an exciting packaging innovation in the grocery trade. In 1902 Ball pioneered the use of the corrugated shipping box for glass containers. The company branched out with acquisitions in this field as well, purchasing the Westside Paper Mill of the United Boxboard Company in Muncie in 1916, the American Strawboard Company plant in Noblesville, Indiana, in 1923, the Eaton, Indiana, Strawboard Mill in 1933 and other smaller facilities.

Its developing prominence in the manufacture of fruit jars was mirrored by Ball's renown in general improvements in the science of food preservation. Years after early machine and food preservation advances, Edmund F. Ball would comment:

"Ball's own research and that sponsored by Ball at university centers also pioneered food preservation technology which is still used today. This early work had great impact on this nation's ability to preserve perishable harvests so that its people could enjoy an ever-increasing standard of living.

"The homemaker's favorable experience with home canned foods led to her acceptance of com-mercially preserved foods. Many well-known food products in grocery markets today started from a favorite recipe of a good cook using Ball jars."

The fortunes and failures that "bigness" means were shared by both the workers and management. When production was up during the years of World War I, salaries were at an all-time high. But after a drastic falling off in sales following the peak production year of 1931, workers' salaries had to be cut. Management concern for workers, however, was a constant at Ball Brothers. For example, workers' wages were upgraded by 15 percent the year following the cut, and raised by another 10 percent the next year. These voluntary wage hikes at Ball Brothers led the *Muncie Star* to comment:

"Probably the biggest factor in the [labor] situation here is the friendly understanding that has existed between the workers and the factory owners. This is due in large measure to the fact that most of Muncie's industries are home-owned and operated directly by their owners. The owners are in direct contact with their workers almost every day in the year, and whenever differences have arisen they have been settled promptly and in a friendly spirit. And local industry has not waited for demands to be made upon it for wage increases as evidenced by the Ball Brothers Company. . . .

"The result is that Muncie has escaped most of the serious labor troubles that have so badly crippled other communities at various times."

Frank described the company spirit when he spoke to his employees at a gathering in 1938, saying, "I have addressed you as fellow workers and I mean just that—we are all working together in one industry."

He meant just what he said. The Ball brothers

tried to prove worthy of the *Star*'s homage. John Ferris, editor emeritus of the newspaper, recently recalled: "To me, all the Balls were down-to-earth people, never ostentatious. They never showed their money. You never saw them wearing diamonds or jewels." Another Muncie newspaperman, columnist Dick Greene of the *Muncie Star*, tells a revealing story of George A. Ball in his later years. Greene, then a reporter, was interested in boarding the Freedom Train when it toured the country in 1947-1948 with important documents in American history.

When it stopped in Muncie, crowds of people queued up in long lines to see the rare exhibits. Being off duty from the paper, Greene sought no special privilege and took his place at the end of the line. "When Mr. Ball came down to the train, of course, he was immediately recognized," Greene relates. "And they said, 'Mr. Ball, you don't have to wait. You can come up here right away.' George said, 'No, thank you.' And he got at the end of the line. That endeared that man to me very much. He could go to the head of the line, but, no, he's a man in his eighties and he'll wait."

Other members of the family were of the same disposition. Editor Ferris remembers of Ed that "everybody liked him. He didn't impose on anybody. He was willing to do anything to help. Out in the factory I've heard more than one employee tell about looking up from his work and there was Ed Ball standing right beside him. Sometimes he'd actually go to work alongside the men, right on the line, rubbing shoulders with them and cracking stories. G. A. was the prankster of the bunch. He enjoyed a good story and fancied playing jokes on close friends. He was a dynamo, always rushing, never walking. He was practically in a strut when he'd go down the street." Frank, says Greene, "was the ruler of the clan, stern and serious. His head was all the way up. Either sitting or standing, he always had his head up." William H. Ball, son of William C., recalls Frank as being "the patriot of valor. He was the leader, a good talker, a very magnificent man." His own father, William, he says, was "a student of literature with a wonderful library of books. His language was at times Shakespearean. Lucius was 'the quiet doctor' of the family."

Miss Elisabeth Ball, seated in the home her father built on the bank of the White River, just where it takes a wide meander on the outskirts of town, reminisced just recently about the site chosen by the five brothers:

"When the Balls came to Muncie everybody who was anybody, as one lady expressed it, lived up on East Washington Street. Mother and Father had come from Buffalo and they were used to wide streets and deep lots and the town seemed very cramped to them and the lots very shallow. So they decided that they would not build on Washington Street and they bought this piece of property that we all lived on and it was a farm. . . . The lawns just went down to the river and they put the street through."

They had built their Muncie plant.

Now they would build their homes.

Chapter 7

"HOW DEAR TO MY HEART"

Muncie was to have a meaning for the Ball brothers which exceeded even the most flamboyant blandishments of the city fathers when they invited Frank to their community.

It was indeed, as they claimed, an ideal place for a company to locate, and for its workers to live. And curiously enough, two literary events of note subsequently were to highlight these two Muncie characteristics. First came the book *Middletown: A Study in American Culture*, by Robert S. Lynd and Helen Merrell Lynd, published in 1929 after what was, for the times, extraordinary research (although it proved to be deceptively shallow in spots).

Middletown was Muncie, although the book never directly revealed its identity. The choice could well have been inspired by Muncie's own Chamber of Commerce, which in the mid-'20s had issued a brochure trumpeting some impressive facts about the community, and stressing its centrality. It was advantageously situated near the precise center of the nation's population at the time, and within a radius of 300 miles of it lived one-fourth of the population of the entire United States. Six railroads led to all markets of the country and to distributing points for foreign trade. Muncie boasted of five interurban lines, and the second largest interurban terminal station in

the state, from which more than one hundred cars arrived and departed every day. The community was proud of the Durant automobile manufactured there, a remote descendant of the original Inter-State in which the Ball family was an investor, and even more proud of the Ball company, "the largest manufactory in the world of fruit jars."

The authors of *Middletown* were drawn to Muncie by its archetypal image as America's big-little town. Clark Wissler, of the American Museum of Natural History, called *Middletown* a "pioneer attempt to deal with a sample American community after the manner of social anthropology. . . . To study ourselves as through the eye of an outsider is the basic difficulty in social science and may be insurmountable, but the authors of this volume have made a serious attempt, by approaching an American community as an anthropologist does a primitive tribe."

This sociologic approach brought forth both harsh and charitable observations from readers and critics, but the authors noted some solid facts about Muncie before reaching their social conclusions. To build a platform for their study of Muncie in the '20s, they rediscovered the Muncie of 1890:

"Middletown, selected in the light of these

considerations from a number of cities visited, is in the East-North-Central group of states that includes Ohio, Indiana, Illinois, Michigan, and Wisconsin. The mean annual temperature is 50.8°F. The highest recorded temperature is 102°F. in July and the lowest -24°F. in January, but such extremes are ordinarily of short duration, and weather below zero is extremely rare. The city was in 1885 an agricultural county seat of some 6,000 persons; by 1890 the population had passed 11,000, and in 1920 it had topped 35,000. This growth has accompanied its evolution into an aggressive industrial city. There is no single controlling industrial plant; three plants on June 30, 1923, had between 1,000 and 2,000 on the payroll, and eight others from 300 to 1,000; glass, metal, and automobile industries predominate. The census of 1890 showed slightly less than 5 percent of the city's population to be foreign-born and less than 4 percent Negroes, as against approximately 2 percent foreign-born in 1920 and nearly 6 percent Negroes; over 81 percent of the population in 1890 and nearly 85 percent in 1920 was native white of native parentage. In the main this study confines itself to the white population, and more particularly to the native whites, who compose 92 percent of the total population.

"The nearest big city, a city under 350,000, is sixty miles away, nearly a two-hour trip by train, with no through hard-surface road for motoring at the time the study was made. It is a long half-day train trip to a larger city. Since the eighties Middletown has been known all over the state as 'a good music town.' Its civic and women's clubs are strong, and practically none of the local artistic life was in 1924 in any way traceable to the, until then, weak normal school on the outskirts.

"Several years later, as abruptly as it had come, the gas departed. By the turn of the century, or shortly thereafter, natural gas for manufacturing purposes was virtually a thing of the past in Middletown. But the city had grown by then to 20,000, and, while industry after industry moved away, a substantial foundation had been laid for the industrial life of the city of today.

"And yet it is easy, peering back at the little city of 1890 through the spectacles of the present, to see in the dust and clatter of its new industrialism a developed industrial culture that did not exist. Crop reports were still printed on the front page of the leading paper in 1890, and the paper carried a daily column of agricultural suggestions headed 'Farm and Garden.' Local retail stores were overgrown country stores swaggering under such names as: 'The Temple of Economy' and 'The Beehive Bazaar.' The young Goliath Industry was still a neighborly sort of fellow."

In such an atmosphere, the Ball company set its industrial roots firmly into the ground of Muncie. Reports of its early success would not make the eye of a modern accountant glitter but Ball could claim an impressive growth considering the nucleus of the business so recently removed from Buffalo.

The authors of the original *Middletown* wrote a sequel in 1937, *Middletown in Transition*, interpreting changes since their first investigations. Their principal new discovery was that the name of Ball had become virtually synonymous with that of Muncie.

Although Robert and Helen Lynd interviewed busboys, bank tellers, and cab drivers, they shied away from personal meetings with members of the Ball family. They nevertheless managed a great many observations on the impact of the family on the community.

The Lynds' description of the family's strong position in the community contained some thinly veiled criticism of the Balls, but it also revealed more mature judgments. In all, an interesting, if only partially accurate, picture of the Ball family emerged. The authors quoted a local citizen who said,

"If I'm out of work, I go to the X plant; if I need money, I go to the X bank. . . . My children go to the X college; when I get sick I go to the X hospital. . . . My wife goes downtown to buy clothes from the X department store. . . ."

And so on. The obvious identity of "X" escaped no one. Other criticisms of the family were more severe. However, the authors always repeated a cautious reminder:

"It cannot be too often reiterated that the X control of Middletown is for the most part unconscious rather than deliberate. People are not, when one gets beyond the immediate army of direct employees, dictated to. . . . The situation is actually much more informal than this. Even within the family a considerable degree of rugged individualism exists. There is a common sense of direction, but no family 'general staff' mapping the strategy. . . ."

What the authors observed and reported was that the Ball family had investments in many aspects of the community. What they failed to stress, however, was that these investments often did not result in a controlling interest in local businesses, and so these firms could usually operate independently of the Ball family. Rather, their investments were considered by the Balls to be a showing of community support through a sharing of the prosperity and good fortune they enjoyed in the fruit jar industry, a further expression of gratitude for the prodigious labor of the Middle-

town citizens. But as the authors noted, "The 'X' family's presence there [in Muncie] was formidable."

Although their success in business became important in national affairs and to Wall Street, the continued Ball presence in Hometown, U. S. A., and the extent of their local investments to bolster the community economy seemed suspicious to many, including the authors of this American classic.

Business Week magazine quite obviously did not share any such suspicions. In May, 1934, it published a retrospective article and was much more open-minded about the Ball family than the book itself had been:

"You could not write a story about the great days of Florence without mentioning the Medicis. Nor can you write a story of industrial Muncie without bringing in Ball Brothers. Five of them started the company which manufactures, among other things, the familiar Mason fruit jar. Two of them, old men now, are still alive. The second generation is active in the management of Ball companies and Ball investments which maintain the family's pre-eminence.

"The Ball family respects the tradition that dollar nobility imposes obligations. They live in Muncie, they take an active part in every move affecting the city.

"The Balls and other local leaders accepted their Depression responsibilities as a matter of course. E. A. (Edmund Arthur) Ball, son of a founder, became head of county and federal relief committees. The family added to the buildings of Ball State Teachers College to give workers jobs. They merchandized their plant products intensively to insure employment (and profits). The Ball Bros. glass plant came over the hump of the

Depression with 1,250 employees. Not all of this was good management. Some of it was luck.

"As hard times bore down, families returned to the thrifty habits of their ancestors. Home canning came back with a rush and the demand for Mason jars grew into a boom. On top of that, along came beer. The Hemingray Glass Co. (makers of insulators such as country boys shoot at along 'phone and telegraph lines) was bought out by Owens-Illinois. They jumped into production of beer bottles.

"Muncie grinned and asked, 'Will Ball Bros. make beer bottles too?'

"The grins were provoked by the Ball family's traditional abhorrence of alcohol in its festive forms. It is an open secret that Ball Bros. did make some beer bottles—rather than see the orders leave the home town.

" 'How did the Balls know they were used for beer?' demands an employee of the plant. 'The items were officially referred to as beverage bottles.'

"The name of this family bobs up again in a look back at the banking crisis. Of more value than worn assurances from a weakening federal government was the announcement during the blackest days that the Ball family fortune stood back of the banks in which it was interested."

We cannot consign *Middletown* to history, however, without noting that the book started something of a literary cult on the subject of Muncie. The city is regularly visited by so many aspiring writers, researchers, columnists, statisticians, and others attempting to go around the *Middletown* track "just one more time," that the people who currently work at the airport are reputedly able to spot a *Middletown* researcher from his tell-tale questions and furtive notations.

There is, predictably, at this writing a further work of research being readied for publication: *Middletown III*, a three-year sociological study done by Brigham Young University and the University of Virginia. Furthermore, in 1979 a team of professional television producers were in Muncie intending to bring forth an updated product to be, in effect, a TV version of *Middletown.* By no coincidence whatever, the media people are intensely interested in the Ball family. The old cycle begins once again, this time with another book and the vehicle of electronic impulse.

Happily for posterity, another author of note was closely connected with Muncie. Emily Kimbrough could report from her personal experiences of the Balls' family life. This gifted writer, now living in New York City, was born in Muncie in 1899 and went on to achieve many literary triumphs. She was graduated from Bryn Mawr, attended the Sorbonne, and was the first fashion editor and the managing editor of the *Ladies' Home Journal.* She wrote more than a dozen successful books, one of them being the best seller *Our Hearts Were Young and Gay*, in collaboration with Cornelia Otis Skinner. She never left Muncie in spirit and one of her most charming books was *How Dear to My Heart*, in which she told of her early years with the children of the Ball family.

It was not just a casual acquaintance. She was one of their little gang, and therein lay the tale of "Minnetrista," where the Ball brothers built their homes side by side, and established their permanent roots in the community.

The name *Minnetrista* was chosen by the two sisters of the Ball brothers, Lucina and Frances. It was coined by joining the Indian word for water *(minne)* with a Middle English spelling of a word meaning an agreed upon meeting place *(trist* or *triste).* The choice of a riverside Indian grounds for their homes was reminiscent of their childhood days along the Niagara River. In Indian times, the land comprising Minnetrista was used to cultivate corn and also served as a sacred burial place. (Elisabeth remembers that when the Balls broke ground on the site, they found bones and Indian artifacts buried there.) Years later the new residents of Muncie gathered there along the shores of the White River for springtime activities such as maypole dances on May Day.

To the visitor today, "Minnetrista," with its stately row of residences, has a lingering message beyond that of an era almost past, of imposing homes, beautifully landscaped and situated with an eye for privacy and beauty. It bluntly says: When those Ball brothers did something, they did it right.

According to newspaper accounts of 1894, Frank, Ed, and George chose "an elegant strip of land just north of the White River" to "construct a boulevard along the riverbank which is 50 feet above the water's edge."

On the morning of June 26, 1894, Louis H. Gibson, a prominent Indianapolis architect, came to Muncie with plans for the Frank C. Ball home. It was to be built partly of frame, colonial style, and to "consist of nineteen rooms equipped with all modern improvements." Construction was done by Muncie builder O. J. Hager. The home took two years to complete. A stone covering and six stately columns were added outside in the early 1900s. Today only the columns stand, somewhat forlorn but still impressive (a fire destroyed the building in 1967). The same architect drew up plans for G. A.'s home, which still stands, in a woodland setting.

Lucius was the third brother to acquire a residence on Minnetrista Boulevard. He purchased a frame farmhouse already constructed on the site and had it turned around to face the boulevard. The home was remodeled, enlarged, and covered with yellow brick in the early 1900s. William, who came to Muncie in 1897 after closing down the sales office left behind in Buffalo, was the fourth brother to build at the site.

Ed, who was the last of the brothers to marry, also was the last to build his home. Miss Elisabeth Ball, who lived on Minnetrista during the time the house was constructed, recalled:

"The land for many years was a vacant lot and we children loved to go over there and play. There were large boulders and we thought we had a secret palace there. When Uncle Ed's house was built, much to my distress, they hauled the stones away. I said to my mother, 'Do you think they are going to take our palace away?'"

They did. But Miss Elisabeth's "secret palace" gave way to what was called the newlyweds' "dream house..., a gracious, impressive, yet not overly elaborate residence," erected between 1905 and 1907 at an estimated cost of $100,000. The house was the most expensive and impressive of all the Minnetrista homes. Ever since, it has been a major credit to its prominent Hoosier architect, Marshall S. Mahurin. The style of the home is reminiscent of English Tudor architecture. It was built with iron, stone, cement, and Indiana oolitic buff limestone. The house featured stained glass windows and doors, as well as a number of curved bay windows. The inside was done in red oak with dark English oil stain.

The actual construction of the home, at least the preliminary work, was done under the supervision of Ed, who, according to his son Edmund

F., was extremely interested in construction, particularly in cement. Construction workers at the time it was built claimed that Ed even "did some of the work and never asked a man to do something that he wouldn't do himself."

Of the five homes, Frank's was known occasionally just as Minnetrista, while Ed's was called Neboshon, an Indian name meaning "by the bend of the river." George's home was known as Oakhurst. Although the names were not generally known, they were used informally by family and friends.

Life on Minnetrista Boulevard was good to the families, especially the children. Here the youngsters' small bodies and large imaginations had room to play. Everybody's children were everybody else's children."

"It was lovely," Mrs. Lucy Owsley, daughter of Frank Ball, remembers. "We used to go down by the river in the spring and hunt for pussy willows, and then sometimes we would go across the river and find wild violets."

The river was the center of the families' activities. Here were the breezy boat rides on Sunday afternoons and skating in the winter on the ice near the Walnut Street bridge. But on some occasions, the river threatened their lives and homes. Although a lot of money was spent for dredging and reinforcing the banks, the river was known sometimes to get out of control.

Behind the homes were the county fairgrounds. Whenever the circus came to town, the children would watch from upstairs windows the big tents being put up and the horses and elephants being led to the river to drink.

Many summer days were spent in Leland, Michigan, near Traverse City, where the families built cottages in the area fronting on Lake Michigan which became known as Indiana Woods. "The days were full of building sand castles," recalls Mrs. Margaret Petty, daughter of Frank Ball. At night the families would build fires on the beach to roast corn or toast marshmallows.

Mrs. Lucy Owsley tells an interesting and amusing story about the families' earliest days at Leland. Lucy, Uncle Lucius, and Aunt Sarah first went up to inspect the area in 1903, being interested in building summer residences. The area around the lake, she said, seemed "most unattractive" and "we were going back and report to Father that this was no place to vacation." However, before they left they discovered a fire that was left smoldering in the woods. When Lucy attempted to help in the efforts to put it out, she fell into the water of the lake nearby. "All my clothes were wet and we had left our bags, except our overnight bags, at Traverse City about 25 miles away." That night the group remained at Leland while waiting for Lucy's clothes to dry. "The next morning someone suggested we go up into what is now called the Indiana Woods." The day proved full of fun and adventure. She says today, "That's how we were then able to report to Father that Leland was not bad after all . . . . Well, of course, now it's full of all our family and others, but when anyone begins to tell about their being old timers and all, I say, 'Now wait a minute. None of you would have been there if it hadn't been that I fell in the water.' "

When the snow fell on Minnetrista Boulevard there were sleigh rides for all the children, and sometimes the bolder ones would hook up their sleds to the backs of the sleighs. Christmas was a special time. Mrs. Owsley recalls F. C. playing Santa Claus to them all. Down at the plant one year, G. A. had special editions of *A Christmas Carol* distributed to one and all.

Throughout the year the children engaged in make-believe "cowboys and Indians" as they climbed on the roof of Ed's home and "shot" out of a stained glass window. They also indulged in occasional pranks. One of their favorites, enjoyed often by the children of Frank and Ed, was to put pennies or even tiny explosive caps on the tracks of neighborhood trolley lines. Recalls Mrs. Margaret Petty, "We would hide to gleefully watch the people jump as the car ran over the caps, making them explode."

Emily Kimbrough remembers spending time with Miss Elisabeth, a great childhood believer in the existence of fairies. She wrote:

"In the spring, when I sometimes went out to spend the day with Elisabeth, we would mix up magic concoctions from leaves and grass and acorns and one or two secret ingredients. Then we would fill little candy boxes, which we had collected and saved all winter, and tie them to the trees all over Betty's place. This was to attract the fairies. Betty went out early every morning before breakfast to look for them. She told her mother one morning that she almost thought she really had seen one and it made a very exciting day.

"The Ball mothers, and some others, including mine, decided to have a teacher for all of us in the playhouse at Frank Ball's home. All the fathers were holding out for a regular public school education. But the mothers said they had tried that, and now they thought they would just try education."

The families hired a friend of theirs, Miss Reba Richey, to instruct the children on the third floor of the Frank Ball home. To this day, the children have fond memories of her. Author Kimbrough described her as a "young, gay, imaginative, tender, and inspired teacher."

On weekends, the children attended Sunday school, which their fathers helped teach. Afterward, the fathers would get together. Mrs. William H. Ball has said, "Sunday, everybody, all the Ball brothers, walked down to Frank's house and had talks. If one was going to buy a new automobile, then they all talked about a new automobile and they all bought the same kind of automobile. This was on Sundays after church."

Alcoholic beverages were not permitted in the Ball homes. Smoking was a vice frowned upon, although a few of the brothers were known to enjoy an occasional cigar. Mrs. Owsley recalls once seeing her father coming home and smoking a cigar before he got to the house. Some of the other children recall the sight of Ed pacing back and forth on the grounds outside his house, even in the coldest weather, to have a cigar which was forbidden by his wife inside the house.

Mrs. Ed Ball lived in the Minnetrista home until her death in 1957. The residence has since been used by Ball State University for social functions and as a place for adult and continuing education. (It reopened in 1977 after extensive renovation and remodeling.)

Mrs. George Ball died in 1958, the last of her generation in the Ball family.

The photographs included in this volume suggested the character of the Ball homes on Minnetrista. They were large, but that was the custom of an era when one could afford it; even middle-class families had servants who "lived in" as part of the household and were often treated like relatives. *Comfortable* rather than *ostentatious* is the word to apply to the furnishings of the Ball homes, but the taste and culture of these remarkable brothers were obvious when one examined the paintings on the wall and the books in their libraries.

G. A. Ball collected books for children simply because he was interested in what advice books were giving young minds of this time. Later on, these books, so casually collected, were joyfully accepted as a gift from Elisabeth by the J. P. Morgan Library, an institution which does not give space on its shelves to inconsequential publications. There were an estimated 30,000 of them! Elisabeth still cherishes today a collection of a vanished art (to modern eyes an oddity, but in reality a rarity in aesthetics), the "fore-edge" book. Such a book contains hand-painted scenes on the outer edges of the pages, minuscule murals almost. When the book is closed, the edges appear to be covered only with a gold surface. But when the book is bent slightly, a scene—such as of Venice—appears as if by sleight of hand. In some of her editions, painted over 125 years ago, there are two scenes on the edges of a single copy: Bend them one way, and there is a picture in delicate color; bend them the other way, and a second replaces the first. It is a lost art, but one of surpassing delicacy and craft.

The Ball homes contained paintings and sculpture which, because of the masters who created them, can only be termed museum pieces. Near Elisabeth's favorite chair is a tiny portrait of a girl, by Gilbert Stuart, who painted George Washington from life. Close to it is a small but vigorously expressed piece of statuary by Frederic Remington, the famous artist of the West.

The last Ball to live on Minnetrista Boulevard is Miss Elisabeth Ball, who tends carefully the beautiful garden still located outside the home her father built. She is a constant reminder of the way things used to be when she grew up there, when the Ball families lined the shady boulevard and cultivated a short strip of Muncie into a veritable Eden. Today the Spring Beauties still bloom, but the horse-drawn carriages and sleighs are gone. The children who used to run freely and wildly between the homes have been replaced by somewhat older Ball State University students and others. Finally, Minnetrista Boulevard, which for 80 years went without the label of street addresses, now has numerals indicating the way for modern Middletowners who are less familiar with the older world.

Chapter 8

APPOINTMENT
IN TOLEDO

The success that came to the Ball family and their company in the first half century of its existence almost inevitably led to further expansion in glass containers. There were two key reasons behind this growth, a phenomenon that would eventually lead the company in an historic move away from its primary function as a fruit jar manufacturer. One lay in the disturbing financial figures that showed a consistent dropping off in fruit jar sales. The second reason rested with the family's reputation for being free-thinking entrepreneurs. Glass products of every kind were becoming increasingly popular, and while containers for home canning would remain the basis for the company for some years to come, there was a promising future in the production of glassware for commercial packaging as well.

It was not the company's strategy to abandon making fruit jars entirely. However, the drop in demand seemed enough to prompt the decision. Sales declined from 186 million jars in 1931 to less than half that amount the following year. The figure hovered near the 75-million-jar figure throughout the 1930s (the start of a decline which ultimately resulted in the first net operating deficit in 1949). The sales total was only 61 million jars in 1936, leading to a net profit figure for the company of $534,281 before taxes, the

lowest figure since 1914. The company reacted by slashing prices of fruit jars, which did help to boost sales substantially, but the result was only temporary. The company's profits soared in 1937 to $1.4 million, but made an incredible drop the following year to only $5,117.

Company executives, spurred by Fred Petty, a son-in-law of F. C., had read the warning signs a few years earlier and knew something must be done. The minutes of the board of directors' meeting of January 8, 1935, read:

"The president (F.C.) called attention to the fact that the demand for domestic fruit jars has greatly declined during the past two or three years owing to the large accumulation of jars in the hands of the consumers and also due to the fact that a large number of jars having the domestic screw thread finish, which will fit the standard Mason Fruit Jar Caps, have been sold to packers for packing various kinds of goods and these jars, when emptied, are being used for domestic household purposes and are, therefore, taking the place of the regular domestic jars sold through the regular channels. A large number of our Muncie glass furnaces are closed down and all of the furnaces at our outside plants are closed owing to this lack of demand for domestic jars. It will be necessary, therefore, if we are to keep up the usual volume of

our glass production, to manufacture bottles and wide-mouth packersware of various kinds.

"This change, it is estimated, will involve the expenditure of more than one million dollars at our Muncie plant and as much more at our outside plants."

A great boost to the Balls' new expansion came when the Great Experiment, Prohibition, failed and sanctions against alcohol, its bottling and its sale, were lifted. Although the Ball family was not known as one favoring alcohol consumption, the Ball Brothers company was one of the businesses to make containers for alcoholic beverages after Prohibition was repealed. Although early sales figures were low (only 1.8 million bottles in 1935), they picked up quickly and were at 53 million bottles just two years later.

While Ball Brothers attempted expansion in glass container sales, it also had to ensure its capacity to manufacture them. To accommodate the new production, at the recommendation of Ed, Ball Brothers considered building and equipping a modern plant to produce jars and bottles at Corpus Christi, Texas. The Corpus Christi plant was brought to the company's attention at the end of 1935 and there is a memorandum in F. C.'s "black book" dated January 3, 1936, that the land could have been purchased and the plant built for $350,000 instead of repairing the Wichita Falls plant, which was later abandoned. But no action was taken. Meanwhile, the other facilities were being revamped with new machines and furnaces. The company then eyed the Three Rivers Glass Company of Texas which had gone into receivership. Ball Brothers made the highest bid for the company, which it purchased in November, 1936, but later sold the bonds back to the company.

While Ball Brothers expanded into new markets, problems continued with its fruit jar business. For years many people were suspicious that the company was forming a trust to protect an alleged "monopoly" of the fruit jar trade. Although the company earlier in the century had joined in some fruit jar "combinations" in an effort to bring down prices and effect better distribution of jars, in fact it never had a monopoly on production and never was involved in forming a trust. But suspicions continued, especially fueled by the trust-busting presidential administrations of Theodore Roosevelt and William Howard Taft, when such industrial giants as the Standard Oil Company under John D. Rockefeller were broken up by the clout hidden in the antitrust laws. During this phase of American history most major industries were scrutinized carefully and certainly the glass industry was not exempt from such a careful watch.

By the 1930s, a system of granting patents for the use of the most sophisticated and modern glassmaking machinery became particularly suspect. A patent war, which had been long fought between two major glass-machine makers, was finally settled, and this resulted in an agreement which would lead, fortunately or unfortunately, to an alleged conspiracy to dictate a master plan of operations for the entire glass industry. The alleged patent monopoly lasted through three decades of the twentieth century, finally culminating in a lawsuit (United States vs. Hartford-Empire, et al.) which would engulf nearly every glass producer of note, including Ball Brothers, and alter the face and future of the industry forever. Specifically, it affected most dramatically Ball Brothers' posture in the fruit jar industry as well as its planned growth in new glass areas. It would lead ultimately to the company's further detachment from the same fruit jar business that had made its name famous throughout the world and would lead also into a bold thrust for survival that would take the company into still more new and exciting ventures.

The early history of the case began in December, 1895, when Michael J. Owens and Edward D. Libbey founded the Toledo Glass Company in order to promote Owens' advances in glassmaking machinery, primarily the continuously rotating suction-charged machine. This firm later became the Owens Bottle Machine Company and finally, in 1929, through a merger with the Illinois Glass Company, became the Owens-Illinois Glass Company of Toledo, Ohio.

Earlier in the twentieth century, the company had granted exclusive rights to Owens' inventions in specific fields and also acquired stock interest in some of the licensed companies. Hazel-Atlas was licensed under the Owens patents to make wide-mouth ware, Thatcher to manufacture milk bottles, Ball Brothers to make fruit jars, and Illinois Glass Company to make narrow-neck ware. The testimony in the case showed the effect of this practice was to put manufacturers not licensed by Owens in the position of developing other processes, of using older methods or of having to get out of this portion of the glass business.

Competitors reacted by extensively financing research into alternative new processes, the result of which proved to be another major advance, the "gob feeder." This machine worked by feeding automatically measured molten glass in the precise amounts needed to form the finished bottle. In the process, automatic shears cut the gobs and dropped them into the molds. What was later to become the Hartford-Empire Company (today the Emhart Corporation) was largely responsible for the introduction of this machine and began licensing and leasing the valuable feeders in the same way Owens once did its suction machines.

A battle between Owens and Hartford continued to the point where, according to a Hartford memorandum, "The industry is pretty well demoralized by competition." However, the two companies continued to engage in extensive litigation, resulting in further confusion and uncertainty until, at last, a series of conferences was arranged in an attempt to settle their differences.

Finally, on April 9, 1924, the two companies ended the dispute and signed a formal agreement.

According to documents produced 15 years later by the government in its court action against the industry, "Owens granted to Hartford an exclusive license to Owens' patents relating to feeders and forming machines, and Owens received the right to use the Hartford patents for the manufacture of glassware." Court papers also were to state, "Immediately after the consummation of this agreement, the two companies set out to obtain control of the remaining feeder patents in the industry, sharing equally the costs and expense between them." Once all opposition to control by the two firms had ended, "there were only four small companies engaged in the manufacture of glassware which were not under license and lease from Hartford, representing less than 4 percent of the total production in the industry" (42 Federal Supplement).

Prior to 1933, Ball Brothers was the largest domestic manufacturer of home-canning jars, but Hartford had never been successful in securing Ball Brothers as a licensee. Ball Brothers was operating well independently on machines and feeders of its own design as well as on the suction machine of Owens, the right to the latter having been previously obtained through the acquisition of the Greenfield Glass Company. In 1933, Ball Brothers' Owens license expired. On March 25 of that year Ball Brothers capitulated after protracted negotiations and entered into its first license of the patents controlled by Hartford-Empire.

At the time the U.S. court suit was filed in 1939, Ball was making 54.5 percent of all fruit jars in the country, Hazel-Atlas 17.6 percent, and Owens 6.4 percent. Kerr Glass Manufacturing Company, one of the four glass manufacturers not licensed by Hartford-Empire, produced 21.5 percent.

The government claimed during the Toledo trial that, as a result of the Hartford-Empire practice, "no new concern could engage in the manufacture of glass containers, except with the consent of Hartford extended through a license to use its machines—feeders, formers, or lehrs." It was also alleged that no new glass manufacturers had entered the industry during the reign of Hartford, although Hartford officials claimed that it had licensed two newcomers.

On December 11, 1939, charges were brought against the glass industry giants using the Hartford patents, in the U.S. District Court for Northern Ohio, Western Division, in Toledo, long a major center of the world's glass production.

Other defendants in the case were Owens-Illinois Glass Company, Hartford-Empire, Ball Brothers, Corning Glass Works, Hazel-Atlas Glass Company, Thatcher Manufacturing Company, and Lynch Corporation (manufacturers of glassmaking machinery). All were charged and eventually found guilty of contracting, combining, and conspiring with various other corporate and individual defendants to restrain trade in violation of Sections 1, 2, and 4 of the Sherman Anti-Trust Act and Section 3 of the Clayton Act. The suit also referred to patents and inventions in regard to machines used in the manufacture of glassware and the fruit jars.

In its defense, Hartford-Empire argued that the patent arrangements increased output and decreased the price of bottles. Bankruptcies in the business, it said, were seldom and technology was moving ahead. Without planned production, it was argued, unemployment, bankruptcies and overproduction would follow.

The trial went on for 112 days and included 3,000 exhibits and testimony and arguments totaling 12,000 pages. Twelve corporations and 101 individuals were involved originally. Among them were George A., Edmund F., and William H. Ball and Fred J. Petty and G. Fred Rieman. Three companies and 40 persons were ultimately dismissed or granted summary judgments, including Frank C. and E. Arthur Ball and Alexander M. Bracken.

Defendants included some of the most important men in the manufacture of glass in those days. Among them were William E. Lewis, Harold Boeschenstein, W. H. Boshart and C. J. Root, all of Owens-Illinois; Armory and Arthur Houghton of Corning Glass; J. Harrison McNash and Walter H. McClure of Hazel-Atlas; W. H. Mandeville of Thatcher Manufacturing; S. G. H. Turner of Turner Glass; Thomas Chandler Werbe of Lynch; F. Goodwin Smith, Roger H. Elred and Karl Peiler of Hartford-Empire; and Charles R. Stevenson and Emory G. Ackerman of Stevenson, Jordon & Harrison, a management consultant firm.

The district court's judgment was handed down in 1942 and it was then appealed to the United States Supreme Court. The final judgment came in 1947 and amended the earlier opinion.

The final decision specified that "a license and lease system may be perfectly legal and just if properly used. However, in the instant case, there has been a deliberate abuse and misuse of that system; and the court believes that there is a semblance of that system remaining. It is the opinion of the court that this entire system must be abolished."

The decision was implemented by a number of restrictions and guidelines that the companies and individual defendants would be obliged to follow, limitations which would be the center of lingering uncertainty and controversy.

Ball Brothers was "perpetually enjoined" from making acquisitions from the other defendants in the case relating to glassware. Also, no acquisitions could be made, without court approval, of the business assets or stock of any competing company. This included the co-defendants' subsidiaries, successors, parent firms, or subsidiaries of a parent firm. Acquisitions could be made of companies other than the defendants for the manufacture of non-container items.

In legal language, Ball Brothers was prohibited from "acquiring, purchasing and holding or acquiring and controlling, directly or indirectly, or through agents, representatives, or nominees, the business of a competing corporation, firm, or individual so engaged (in the manufacture and sale of glassware or in the manufacture or distribution of machinery used in the manufacture of glassware), unless any such acquisition is approved by the court."

Ball Brothers was directed specifically to divest itself of all assets and business at its facility at Three Rivers, Texas, which it had acquired only in 1936.

One lawyer involved commented professionally on the case: "From what I know of Ball Brothers' business—which, in the glass field, seems to embrace only a limited area in the container field—it seems to me that, purely from a legal standpoint, it has opportunity for expansion in the glassware field generally, and even in the container portion."

Product or territorial diversification and expansion without any limitation was permissible if Ball Brothers built its own capacity. Furthermore, acquisitions of any kind could be made from any corporation engaged solely in business outside the United States and its possessions. Acquisitions could also be made of the business, assets, or stock of a noncompetitive firm. According to one opinion at the time, these acquisitions "in some cases would probably be sufficiently clear for counsel to take the responsibility of determining that no competition exists; in others it may be advisable to obtain a court determination."

After almost 50 years of innovation and growth in the manufacture and distribution of fruit jars, Ball Brothers was not as sure of its opportunities as the optimistic legal comments suggested. Years would be spent studying the decision and exploring what avenues for growth were still open. It was decided, finally, that if Ball Brothers was to continue to thrive, it would have to diversify.

It was a bold attitude. There would be the loss of the strong single-product corporate identity the firm was accustomed to, but it was a stance that would provide the company the opportunity to prove itself as capable in other ventures. This was 1945; the world was on the verge of an atomic age, computers, and an era of high technology. Perhaps the next generation could establish Ball Brothers' leadership in industries yet unknown.

The concept of diversification thus thrust upon Ball Brothers by the courts was not exactly novel in the company's planning. Years before the legal charges against the firm were made, then-Assistant Secretary Edmund F. Ball recommended that the company set up a program looking toward diversification of manufacturing activity. He said in 1936 that the making of fruit jars, Ball's chief product to date, would continue, but that entry into other fields could provide for more steady and possibly increased employment.

To that time, Ball Brothers had moved in many directions which were in some way usually related to fruit jars. By the 1930s Ball Brothers was operating a rubber department for the manufacture of the rings used in sealing the jars. It expanded its zinc production throughout these years and completely remodeled its Muncie mill in 1934. The company had purchased a number of paper mills, among them the Eaton Paper Mill in 1935. In 1926, Ball Brothers entered the manufacture of shells for "B" and "C" batteries for use in radio and flashlights. In 1935 it entered the pressure-cooker field. However, in these areas they found themselves competing with important customers and withdrew.

Between the years 1924 and 1931, the future of Ball Brothers might well have been entirely altered had any one of a series of proposed mergers materialized. There were serious talks over those years involving at least five purchase or merger proposals.

They came at a time contemporary with or shortly after the acquisition of the three Schram plants in 1925, a move which made Ball Brothers a more attractive company with which to merge. This was enhanced by the purchase in 1929 of the Pine Glass Corporation of Okmulgee, Oklahoma, at a cost of $590,000 for the fixed assets.

During negotiations for one of the proposed mergers, talks apparently went smoothly until a proposed director of the new firm offered the remark that everyone seemed to be happy and satisfied with the terms. Whereupon the attorney for Illinois Glass Company said, "We will be unless the Federal Trade Commission gets after us."

This off-hand remark seemed to have planted a seed of doubt in the mind of Frank C. Ball and he began to have second thoughts on the merger. Upon further examination, it began to look as though the problem might well be that the Illinois Company's previous acquisitions of the capital stock of other companies had drawn complaints from the Federal Trade Commission.

F. C. slowed down the negotiations and it was well that he did. Subsequent discussion convinced him that the Illinois Glass Company planned a corporate setup which would make it possible for them to ignore the Ball people in management. It became obvious that, far from being a merger, the Illinois Company was contemplating bringing about dissolution of Ball Brothers interests in the combined enterprise as soon as possible. It cost Ball Brothers some $200,000 in expenses to withdraw from the negotiations, but withdraw it did. Some inference of the attitude on the part of Illinois may be drawn from a confidential memo circulated to certain members of the company by William E. Levis, vice president and general manager and a grandson of one of the founders.

A copy of this memo was sent to F. C. and he could fairly infer from the wording that he had been prudent in withdrawing from the merger, since Mr. Levis—not particularly to be castigated for so doing—clearly had in mind the well being of Illinois and not very much concern for Ball Brothers.

In 1929 Ball Brothers company was approached by partners of J. P. Morgan & Company and of Swarthout & Appenzeller with the suggestion that they be given permission to put Ball Brothers on the market. The first company mentioned as a buyer, according to G. A.'s notes, was the Gold Dust Corporation. Also indicated in G. A.'s notes was the fact that Mr. Morrow, president of Gold Dust, was something of a horse trader whose opening tactic was to downgrade Ball Brothers to its own chief executive—its plant investment was too heavy in proportion to sales, home canning was decreasing, one or two plants should be sold, etc. Nonetheless, Gold Dust offered $15 million in cash for the business. Ball Brothers was not interested in such an all-cash deal, and the offer was not followed up.

Morgan & Company next came up with a proposition from two unnamed companies which offered Ball Brothers $18 million, part of which, apparently, would take the form of an exchange of stock. G. A. stated that he could hardly be interested in anonymous investors, and that no consideration would be given to any such veiled offer. Morgan then divulged that one of the would-be purchasers was Anchor Cap & Closure and "a glass company." G. A. surmised that the latter was Capstan Glass Company of Connellsville, Pennsylvania. His concluding note on the negotiations was crisp: "We told them that we didn't think well of this matter and did not want to go further and departed."

The next contemplated merger was to be with the Hazel-Atlas Glass Company, first considered in 1928. G. A. Ball originally had been approached by the Guaranty Trust Company in New York on behalf of Hazel-Atlas, but it apparently was at this stage only something of a "feeler" since, although a representative of Hazel-Atlas was present on the premises, he declined to lunch with G.A. and the bankers on the grounds that he was too new with Hazel-Atlas "to be in a position of talking." However there were others who could talk, and meetings were subsequently held in Pittsburgh, New York, Atlantic City, Muncie and Anderson, Indiana. G. A. was interested enough to write the Hazel-Atlas people on April 6, 1929: "We are still giving consideration to both plans you suggested." The proposal was still alive as late as May, 1931. In the end, however, all came to naught. It would even appear that the proposed plans for a merger were agreeable to Ball Brothers, but were never accepted by Hazel-Atlas, and the matter was dropped.

When the anti-trust action by the government came to a climax, Ball Brothers was in the midst of many active corporate considerations. In retrospect, the momentous court decision of 1947 may well have been a fortuitous event for the Ball company, but at the time it was an unsettling and discouraging obstacle to what had been, up until then, a remarkable achievement in growth and company expansion in the glass industry. At this critical time in the company history, the Ball brothers clearly showed their determination to remain in charge of a changing and even troubled business despite a court decision that drew a sharp dividing line between the days of emphasis on fruit jar production and the era of a diversified corporation.

Chapter 9

"ALARMS AND EXCURSIONS"

No story about Ball Corporation would be complete without taking notice of the various ventures entered into by the brothers, bearing no direct relationship to the fruit jar business itself.

Family archives suggest that these entrepreneurial excursions took many forms, of which few records seem to have survived. In the case of George A. Ball, some prima facie evidence is a well-wrapped collection of stock certificates (preserved as mementoes) indicating a propensity to branch out wherever his curiosity took him. He was known to other members of the family as being the closest thing to a "plunger," in the popular sense. The story goes that he went out one day and bought a private, defunct Muncie college, which—now having it on their hands—the family proceeded to develop into what is now Ball State University. Actually, this is quite an apocryphal version—the entire Ball family, especially F. C., had been interested for many years in establishing a college in Muncie and the "purchase" was far from a whim at the moment. However, traditionally, G. A., who once bought a railroad in Alaska, also bought a college because he won out in close bidding for an institution on its last legs financially.

An interesting thing about G. A.'s character

(the more so if we accept his reputation as one who enjoyed taking chances) is a contrasting part of his personality, which was abstemious in the extreme. The contents of his various desks have been dutifully preserved in his memory, and one cannot resist a slight smile at some of his memorabilia. Box after box is filled with pencils no more than an inch long, too short to use but obviously with enough lead left in them for G. A. to feel it would be wasteful to discard. There are also penknives with broken blades, matchfolders bereft of matches, out-of-date calendars preserved perhaps because of an interesting bit of decoration on them. There are other bits and pieces of metal of unknown origin, evidently saved for some future day when they might come in handy during a sudden emergency.

One nods affectionately at the same parsimonious trait exhibited as G. A. is photographed in his office while being interviewed during his term as chairman of the firm by a newspaperman seeking some insights from the highly articulate and quotable multimillionaire. G. A. sits there in the photograph, kindly and thoughtfully answering questions, with his legs comfortably crossed, exposing the fact that the shoe facing the camera has a hole in its bottom and is obviously in need of re-soling or retirement. As one friend said of G. A., "He'll carry his own suitcase to save a dime and then buy the fairgrounds for his daughter to plant flowers in."

No doubt of it, G.A.'s sense of economy was extended beyond his personal habits and into the operation of the business, where "a dollar saved is a dollar earned" had real application. But the world was to learn of G. A.'s canny sense of values in a highly spectacular excursion into big-time railroading which was actually thrust upon him in a most unusual manner.

There were five brothers Ball, but four Ball brothers and their individual personalities were little known to the outside world.

G. A. was always a shy man who nevertheless began to emerge as a distinct character after the death of older brothers Ed and William. Between 1932 and 1937, he was Republican national committeeman from Indiana at the same time the family was keeping a relatively low profile back home. A letter survives in the archives, from Herbert Hoover, who thanks George for his support despite his loss to Franklin Roosevelt in the 1932 presidential election. Also in the archives are a number of photographs depicting George with leading G.O.P. figures of his day such as Alf Landon and Wendell Willkie. Other pictures show George with top film stars like Rosalind Russell and Robert Montgomery. Edmund F. Ball remembers G. A. introducing Alf Landon at the Indianapolis Coliseum before a large crowd during the presidential campaign of 1936. He took his nephew Edmund and Fred Petty with him to Elwood, Indiana, to hear his friend Wendell Willkie, the Republican candidate for the presidency, make his famous "cornfield" acceptance speech in 1940.

When G. A. drove an automobile, he drove it fast. When he walked, he went quickly, often to the wonder of other people on the sidewalk. He loved practical jokes, even of the most schoolboyish kind, such as putting tacks on other people's chairs. G. A. in fact loved jokes even when they backfired, such as the one he told on himself. It was at a time when he was still on the board of Midland Marine Bank and was making trips frequently between Muncie and Buffalo to attend meetings. On one such occasion, while in his hotel in Buffalo, he remembered a nearby friend he had not seen for a number of years. On the spur of the moment he called him on the telephone. The man was delighted to hear from his old friend, learn about his health, and how everything was going in Muncie. They enjoyed their telephone conversation so much that they made arrangements to meet at the hotel and extend the visit. Impulsively and in a joking manner, G. A. took a notion to say: "And by the way, bring a little money. I might want to borrow some from you to pay my hotel bill and get back to Muncie." He recalled that he sensed immediately he must have said the wrong thing, as he detected a distinct cooling in the tone of his friend's voice. He went down to the lobby as arranged at the appointed hour and waited to no avail. His friend never showed up.

G. A. loved fun and thrills. His frequent plunges into non-glass investments caused his family concern. Stories are still told of attempts to restrain him whenever he had a "hunch" to invest. It was this financial daring, together with his penchant for publicity, which brought on his most famous exploit: acquiring the tangled Van Sweringen railroad "empire."

The situation was set up by the effects of the Great Depression on the Van Sweringen brothers of Cleveland, Ohio, owners of 28,000 miles of railroad.

The Van Sweringen brothers, Oris P. and Mantis J., were modern operators who still had something of the swashbuckling manner of the old railroaders clinging to them. It was their dream to create one transcontinental railroad. They bought railroad after railroad, pyramiding their transactions during the '20s in the years just before the stock market crash. In '29, the state of the Van Sweringen affairs was conservatively described as "an unprecedented, historical, age-defying and entangling mess."

The two brothers controlled a corporate empire which included such great railroads as the Chesapeake & Ohio, the New York, Chicago and St. Louis (also known as the Nickel Plate), the Pere Marquette, the Erie, and the Wheeling & Lake Erie.

Some holdings are more difficult to name specifically. The Van Sweringens had complicated financial set-ups for these companies, and the involvement of courts, creditors and receivers entangled the issue of ownership further.

The Alleghany Corporation, for instance, controlled the Missouri Pacific and through it the Chicago and Eastern Illinois, the Denver and Rio Grande Western, the Texas and Pacific, and the International Great Northern. Through the Rio Grande, it held a half interest in the Western Pacific. (However, the Missouri Pacific was in receivership during the 1930s and was therefore actually controlled by its creditors through the courts and a receiver.)

The Van Sweringen interests also included steamship lines, trucking companies, coal mines, trolley lines, and even a peach orchard in Texas. Among these interests were the famous Halle's Department Stores and the Terminal Tower in

Cleveland, and the palatial Greenbrier resort hotel in Virginia. But all their investments were near bankruptcy. Ultimately the Van Sweringens could not refinance their $40-million loan from J. P. Morgan & Co. and consequently the bankers planned to sell the Van Sweringen collateral for the loan at auction to the highest bidder. No question, the Van Sweringens were in real trouble. Although they turned everywhere for help, they were seemingly at the end of the road.

A friend of theirs, George Ashley Tomlinson, of Cleveland, principal owner of a fleet of Great Lake freight carriers in which the Ball family had invested, was married to a niece of Mrs. Lucius Ball. G. A. knew the Van Sweringens "not particularly well," but he had been asked to join one or two of their directorates. Beyond that, he was not a man to turn down a call for help without a hearing. As one of his friends once said of G. A.: "Nothing is too small to interest George Ball, and nothing is big enough to scare him."

So when Tomlinson telephoned G. A. and said: "The Van Sweringen brothers are in a hole. They want to talk to you," G. A. agreed. He had more than curiosity on his mind, though. As he later explained to *Nation's Business* magazine, after the "Van Sweringen affair" had made countrywide newspaper headlines and caused a furor in Washington and Wall Street (as well it might, considering that the Van Sweringen properties had a potential value of $3 billion):

"If those properties were sold in the open market some of them might go at the eighth bidder— the lowest allowed on the stock exchange. They might fall into the hands of crooks and amateurs and chiselers and be destroyed. Something like 25,000 people have been working for the Van Sweringens in the city of Cleveland alone. No one

can say how many of them would find themselves out of work. No one could say how many other people elsewhere would lose their jobs. If that huge, tangled, complicated mass of properties were to be overturned, the ruin might embrace the nation itself. We were just then beginning to see a chance of regaining our vanished normality. Another smash might have set us back—who knows for how long? Who can say how much harm might have been done?"

At the time, however, G. A. kept his own counsel concerning the Van Sweringens, and merely consented to see them and hear their plea. On December 24, 1936, he wrote, apparently for his own record, a retrospective memorandum which the newspapers of his time might have given a great deal to get into their hands, since it revealed his private thoughts on something which had by then become a subject of national interest. The note further reveals G. A.'s terse but dramatic style of writing. It contains a highly dramatic moment when he came to a decision, in his own way, on the Van Sweringens' plight:

"On August 10th [1935] I received from Cleveland a telephone call from George A. Tomlinson, a relative of the Ball family and in whose fleet of steamers on the Great Lakes many members of the Ball families have been interested for a number of years. Mr. Tomlinson said that the Van Sweringens had a matter they would like to talk over with us and when could it be done. It was arranged that they come to Muncie the following day, Sunday. They did come, O. P. and M. J. Van Sweringen and Mr. Tomlinson, arriving by automobile about two o'clock, Mr. M. J. having done the driving of the car.

"Mr. O. P. promptly began a recital of his situation and practically gave the story of their lives—

their start in real estate and leading into and expanding in the railroads and ending with a statement of his own immediate critical situation which was that he was owing banks some forty million dollars. The banks were going to sell the collateral and he was looking for someone who would buy the securities and hold the system together instead of allowing it to be dismembered and the benefits of the years of work and assembling and profits to be dissipated. He explained some of the high spots leading to his difficulty and present embarrassment. He told of his having used several millions of his personal fortune in an attempt to save one or two companies of which he was only a part owner. His story was very graphic. As he put the matter to us, I thought that we were confronted with an economic, social, financial, political problem and that it would be a calamity if the great system which had been created would be allowed to be dismembered, one property sold here, another there and others somewhere else; all the headway made in grouping these various properties, partly mapped out by the government, would be lost.

"The financial outlook was not particularly good just at that time and it seemed to me that most business and financial men were a little jittery. I felt that any such happening as throwing on the market such a large lot of securities and the kind they were would make people more jittery and would have a very harmful effect and cause a slackening up of the then slowly returning of confidence. I felt that anyone to whom such a situation was presented had a moral duty to try to do something to be of what help he could.

"We asked what amount of money might be required to handle the matter. Mr. Van Sweringen stated he thought that some two or two and one-half million with such additional as could be secured by bank loan on the securities purchased. I asked what he would do if after giving the thing some consideration we should decide to do nothing. He replied that he had no other place to go. He had thought of many plans and was now at the end that if we did not do it, he was through and would have to fold up. We suggested many individuals and firms for him to try. There seemed to be some barrier in every case, some reason why he thought there would be no favorable reception.

"Mr. Tomlinson's plan was to leave Muncie that evening for Colorado instead of returning to Cleveland. We had talked about two hours and we then told Messrs. Van Sweringen that we, Mr. Tomlinson and I, would talk the matter over that evening and let them have some word in a few days but suggested that if he had any other place or person to get the necessary money from, he should not neglect that other avenue nor lose contact with any such source for, of course, we might decide unfavorably. He replied, 'Mr. Ball, as I told you before, we have no other avenue for the money. We have thought of everything and everybody and if you men don't do it, we are stopped dead.' As he went out the door, he had a commitment from me but he did not know it as I did not tell him then. That evening, Mr. Tomlinson and I talked the matter over. We had both a high opinion of the properties and a high appreciation of the managing organization having a hand in the running of the properties and we both felt confident that with proper handling, the situation could be worked out. We decided that we would try to interest one other party and each supply one-third of the money necessary.

"During the next few days there was concluded the arrangement resulting in the formation of Mid-America Corporation, which made the purchase of the securities at the auction."

On September 30, 1935, G. A. and Tomlinson accompanied the Van Sweringen brothers to the auction. Before the meeting, G. A. and Tomlinson formed the Mid-America Corporation as a holding company, according to the wishes of the brothers. G. A. and Tomlinson had previously agreed to come up with $2 million between them and then borrowed an additional $1 million from the bank. At the meeting, G. A. bid $3,121,000 for control of the Alleghany Corporation, top holding company of the Van Sweringen financial pyramid, and thereby bought back the firm for the brothers. A short account in the New York Journal American, a William Randolph Hearst newspaper, described a sidelight at the auction:

"Reporters saw a dignified white-mustached man sitting beside O. P. Van Sweringen. . . .

" 'Who is that sitting over there beside O. P.?' asked one reporter.

" 'Maybe it's Santa Claus,' another newspaperman suggested.

" After the auction, it was found the unidentified man was Santa Claus."

The deal G. A. made with the Van Sweringens was that they would remain in control of the railroads. He also gave them the option permitting them to repurchase control when they were in a position to do so. Certainly it seemed like an act of St. Nicholas.

G.A. said much later of the event: "I was not moved by the thought of profit."

But something happened shortly thereafter that put G. A. in an unforeseen position. He had bought out Tomlinson's share in the deal, so when the Van Sweringen brothers both died suddenly within months of each other, apparently exhausted by the effort to save their holdings, George A. Ball found himself owner of all those railroads and other properties the Van Sweringens had amassed. Overnight he became a national figure and virtually a household name. *The Saturday Evening Post* featured an article about him, "Mr. Ball Takes the Trains," and *Collier's* told the story under the headline "Bet You an Empire." A nationally syndicated three-part article described him as a "quiet, retiring man of great affairs" and newsmen from across the country swamped him wherever he went.

But, typically, G. A.'s new prominence made no significant change in his personality. His daughter, Miss Elisabeth Ball, recalls that when her father came home after buying the Van Sweringen "empire" he merely remarked to his family at the dinner table that evening that he had just bought a peach orchard in Texas he thought might please them. He didn't mention anything else he bought along with the orchard from the Van Sweringens—like a railroad empire.

But while George was becoming one of the most celebrated men in national business affairs, other things were brewing to test his mettle. Shortly after the death of the last Van Sweringen brother, Democratic Senator Burton K. Wheeler of Montana called his railway investigation committee together and summoned George A. Ball to testify. G. A. rose admirably to the occasion and admitted quite frankly that he knew little about the empire of which he then found himself in control. Reporters present were impressed by his honest and candid discussion of himself and past business dealings. Furthermore, G. A. kept the mood of the hearings light, every now and then interjecting a remark to charm the eastern establishment. His quiet humor came to light during the hearings with Senator Wheeler. At one point, the famous senator recalled that George was Republican national committeeman from Indiana, and more or less twitted him on the fact that he had picked Herbert Fitzpatrick to operate the Van Sweringen system for him, even though Fitzpatrick was a former Democratic national committeman from West Virginia. G. A. admitted to that. The senator then said, "I was wondering if you couldn't select the Socialist national committeeman from some state, and you would then get them coming and going."

G. A. played it straight and didn't bat an eye. "I would be very glad to talk that over with you," he avowed to the nonplused senator.

But though G. A held tight to the railroads and peach orchards he showed no real interest in their day-to-day operations. He chose other men to operate the companies and went back to Muncie and the family and the fruit jar business that was his life. Reporters and celebrities followed him there to visit and to speculate. One writer for *The Saturday Evening Post* seemed struck by the fact that at the Ball company G. A. was only a vice president, and that his office "looks just about the way it has looked for the past ten or fifteen years.... From his office window Mr. Ball looks out on no empire of railroads. Across Macedonia Avenue a sign on a murky window says: HAIRCUT 25 cents."

After only two brief years of owning the "empire," unpredictable G. A. was true to form and once again startled the business world. In March, 1937, he sent out a memorandum announcing the establishment of the George and Frances Ball Foundation for "educational, religious and charitable purposes," similar in many ways to the Ball Brothers Foundation. The business community was jolted when one month later G. A. transferred over to his new foundation all common stock of the Mid-America Corporation. Shortly thereafter, the George and Frances Ball Foundation sold the stock for $6,375,000 to Robert Young, Frank E. Kolbe, New York investment brokers, and Allan Kirby, of Wilkes Barre, Pennsylvania, an heir to the Woolworth fortune. Later, G. A. faced a $5 million court suit alleging "transactions on the New York Stock Exchange known in financial circles as 'wash sales' for the purpose of creating both actual and apparent active trading," but an out-of-court settlement ended the case.

G. A.'s excursion into the Van Sweringen empire was by no means a singular occurrence in his financial life. His name appeared in connection with many business ventures which had nothing to do with Ball Brothers. He backed a young man named Clyde E. Whitehall in the Banner Furniture Company, which enterprise spread from Muncie to Indianapolis and Columbus, Indiana, and Toledo, Ohio. He invested in Intertype Corporation, makers of typesetting machines, and was a director of Dictaphone Corporation.

He bought up a bankrupt company in Danville, Illinois, devoted to the manufacture of paper from corn stalks and arranged for Ball Brothers and J & H Sloane (now W & J Sloane, Inc.) of New York to run it jointly and process hemp for rugs. He invested in General Household Utilities Company and the Durham Manufacturing Company. The latter firm was a Muncie company, now moved out of the city, which had manufactured fireless cookers and other household appliances. Another adventure was a mineral mining company in Kentucky. A partial list of the companies in which he was a director includes Borg-Warner Corporation, Kuhner Packing Company, Great Lakes Portland Cement Company, the Nickel Plate railroad, and Merchants National Bank of Muncie.

G. A. served the company as its president for only five years after Frank's death in 1943, retiring in favor of his nephew Edmund F. Ball in 1948. He then became chairman of the board, but his influence, like Frank's, was a continuing factor since the company's beginnings.

Although G. A. was unrivaled by the other brothers in his penchant to explore new businesses, he was by no means the only brother to do so. F. C., W. C., and E. B., too, had a flair for reaching into the unknown and undertaking new risks. Three-inch headlines in red ink and eight-column-wide photographs of the three brothers on the front page of the *Pueblo* (Colorado) *Star-Journal* dated January, 1908, trumpet out a triumph in civil engineering:

"Ball Brothers complete fountain underflow system and water is now at city's gates."

The Ball brothers were identified not with fruit jars but as F. C., president; W. C., secretary; and E. B., vice president of the Pueblo Water Supply & Power Co.

The text of the long article in that edition, also headlined in red ink—"Business-Like Methods Applied to Work of Construction Accomplish Results"—tells the story. "Representative men of Muncie, aided by competent local superintendent and engineer, take up task of completing the underflow system and water is now at Pueblo's doors."

In brief, the project had been started in 1902 and $600,000 expended on a plan to provide a dependable water supply for Pueblo from the nearby Florence River. A company had been formed to accomplish this, but it seemingly never could get the job done. And in truth, the challenge was no small one. The proposal was to build two three-acre reservoirs, with a storage capacity of 41 million gallons, good enough for a seven days' supply of water for the community. But local interests were unsuccessful and at length were compelled to turn to investors in the other parts of the country in an attempt to finish the job. As it happened, the Ball brothers had put some money in the project, and when it became clear that in order to get the water running they'd better do it themselves, they took over in Pueblo. As the paper opened:

"It requires but little more than a smooth tongue and sharp pencil to outline a project on paper, and portray in high-sounding phrases the possibilities of the project, but as everyone knows who has had practical experience, it is quite another matter to be able to produce the actual claimed results."

So much for the local company. The newspaper wrapped it up with:

"In this day of accomplishment, ordinary achievements come so thick and fast that little notice, other than to chronicle the event when it transpires, is taken of what men do, but when projects, which, before the hands of brainy men evolve their solution, are considered impossible and so regarded by the lay mind, are finally accomplished and pointed to as finished, then it is that more than passing notice should be given to all such undertakings."

The *Pueblo Star-Journal* thus hailed the three brothers from faraway Indiana as benefactors of their city. It was once more a somewhat prophetic adventure on the part of the Ball brothers. The brothers had again demonstrated a wide and varied outlook beyond the threshold of their home business, one they would maintain throughout their lives and pass on to the next generation.

A curious aspect of the Pueblo episode was that the brothers never were paid for their work due to a dispute over the quality of the water. The absence of payment amuses Ball officials today who recall that it used to be the responsibility of any newcomer in the firm to try to collect on the debt.

About 70 years later, Ball Corporation was to enter into a somewhat similar adventure, that of providing water in an area where it had never been available before. The distance this time was far greater than Muncie to Colorado—all the way to Kufra, Libya, North Africa. In that remote country the company started by the brothers who brought water to Pueblo this time accomplished the greening of the Sahara desert by bringing up water from 5,000 feet below the sands and creating in the wasteland, by means of center-pivot irrigation, waving fields of wheat and alfalfa where, from time immemorial, there had been only the plodding of an occasional camel. The immensity of this modern adventure might well have astonished the original Ball brothers, could they have observed it. Certainly, it would never have daunted them, because, in one way at least, the other brothers resembled George quite closely: Nothing was too big to scare them. It was to remain a fundamental characteristic of their company.

Chapter 10

BENEFICENCE

At a time when the Ball family name was often in the national headlines, resulting from its fruit jar fame, G. A.'s outside financial ventures or deeds of daring such as the Pueblo water project, the Ball clan back home in Muncie was making substantial but comparatively unperceived progress on many cultural fronts.

Over the years, much of the fortune amassed from the growth of the family's original fruit jar business had been turned over directly to community and state philanthropies as expressions of genuine regard for local citizens who had supported the Balls from the very first days of immigration from Buffalo. The extent of these benefactions spanned an array of diverse social services, reflecting the family's longstanding interest in areas such as health, religion, history and education. Taken together, the individual and collective endowments and largesse of the family have done much to perpetuate the name "Ball" beyond the point when it first became a famous trademark on the front of a fruit jar.

The inspiration of Maria Polly Bingham Ball, herself a schoolteacher, and the example of Dr. George Harvey Ball, a pillar of strength behind the first business efforts of the five Ball brothers, impressed upon them at an early date the value of education. In fact, that first $200 loan from Dr. Ball which put the five brothers in business was repaid many times over in the form of continuous financial support for Keuka College. The presi-

dent's residence, close to the shore, must be one of the major emoluments of that position. It is a stately and handsome formal structure, almost a mansion, and bears the name "Lucina" after the older sister of the five brothers.

The Ball family's backing of Keuka College is but a part of their continuing interest in a number of educational institutions. Ed Ball is a loyal son of Wabash College, and is a trustee. Hanover College, where William H. Ball was a trustee, receives family benefactions, as does Hillsdale, where Dr. George Harvey Ball was co-founder and sister Frances met her husband-to-be, William J. Mauck, when she was an undergraduate and he a teacher. Later on, he became president of Hillsdale, as did their son Willfred. The records of the company show repeated financial support of the college. Indiana, Purdue, Taylor, and DePauw universities and Earlham and Anderson colleges also have enjoyed support of the Ball family and company throughout the years.

Probably the most notable of the family's benefactions is what is now Ball State University. Today the school's enrollment is 17,500 students and it is one of the largest employers in Muncie. But the school had humble beginnings. When the authors of *Middletown* first came to Muncie in the 1920s they observed, "In the spring of 1925, its [the college's] impact on the town, other than its increment to local trade, was practically nil. It was an inconspicuous institution out on the edge of the cornfields, on the margin of the city's consciousness."

But it was on the Ball family's consciousness well before 1925.

Late in the nineteenth century, members of the Ball family, like other business leaders in Muncie, were interested in seeing a college started nearby. Not only would the school be a prestigious addition to the community character, it would mean that young adults of the area could continue their education past high school without having to leave their home town.

But early efforts to get the school started, most often spearheaded by a member of the Ball family, proved unsuccessful. Four times the school fell victim to poor economics and had to be closed down. Finally, in September, 1917, a court ordered that the property be sold at public auction. Only two bids were made for the school. Ball's agent gave what was considered "the best bid" and the brothers were granted possession of the property. Within six months they increased it in size by 64 acres and promised to add two more buildings. Then, in 1918, the real start of what would become Ball State University emerged. Having been accepted by the state of Indiana, the school once again opened, this time for good, and was known as the Eastern Division of Indiana State Normal School in Terre Haute.

Soon after the school's first graduation ceremony, Indiana Governor James P. Goodrich unveiled a bronze plaque commemorating the Ball brothers' gift of the school. F. C.'s response was brief. "We are glad of the opportunity to turn an institution of this kind over to the state," he said. "The good it will accomplish in the coming years will more than repay us for what we have donated."

But the family's generosity toward the school did not stop at that point. The scope of its contributions, once the college was at last on its feet, is moving and impressive. Besides outright cash gifts of support, a number of buildings were constructed with Ball family or company money. Among them is the school's gymnasium completed in 1925, the year E. B. Ball died. Ed had taken a special interest in the construction of this facility and even helped in the pouring and mixing of concrete. A women's residence hall was built in 1927 in memory of Lucina Ball. A men's residence hall was built in 1938 with funds from the Ball Brothers Foundation and was named in memory of F. C.'s son, Frank Elliott Ball, who had been killed in an airplane accident.

The Arts Building, considered by many to be one of the most beautiful structures on campus, was built largely with Ball family money. Although built with public funds, a library honoring Alexander M. Bracken, the school's longtime trustee and later president of the board of trustees, was built most recently. Other examples of the family presence on campus include Frank C.'s and E. Arthur Ball's art collection on permanent loan in the Arts Building and the George and Frances Ball Distinguished Professorship in business.

The school reached university status in 1965 and today is a state and national leader among educational institutions. The univeristy has continued to grow and now has five colleges and four associated schools and continues to receive support from Ball foundations.

Health was another area in which the brothers took great interest, doubtless a result of brother Lucius' work in the field. Edmund B. served as a trustee on the board of the Muncie Home Hospital, predecessor of Ball Memorial Hospital. When overcrowding caused officials to consider enlarging the building, Ed offered an alternative. He said that if the state legislature passed a bill authorizing Delaware County to build and maintain a new hospital, he would see to it that the necessary funds were provided for the entire project. The bill passed and Ball State Teachers College provided the grounds for the new hospital. The project included the construction of four buildings at a total expense of more than $2 million. Ball Hospital was opened August 4, 1929.

The Ball family was also one of the principal backers of a plan in Indianapolis to build a hospital for the special care of children. F. C. and some friends came to the conclusion that the hospital was a necessary project and that it should be a memorial at the same time to Hoosier poet James Whitcomb Riley, whose fondness for children shows through all of his writings. Plans to build the hospital were interrupted by the First World War, but shortly after the war, F. C. and other leaders in the state organized a campaign to raise the necessary funds for the building. F. C. personally was also a major contributor to the Riley Home for Nurses, now called the Ball Residence, located on the campus of Indiana University-Purdue University, Indianapolis.

One of the most interesting public contributions made by the Ball family has an unusual historical significance. The land where Abraham Lincoln romped as a boy in Indiana between the ages of 7 and 21 (F. C. called them "the most impressionable years of his life") was in poor condition. A campaign was launched to raise $1 million for an appropriate memorial at Lincoln City, Indiana. The Ball brothers pledged $25,000 to the cause, but when the movement failed, the brothers used their pledge money to make the entire purchase of the land on which the home was located. In his Memoirs, Frank wrote about the property shortly after the purchase:

"On this property was a schoolhouse, church, small hotel, and several small residences. These were all moved off and the grounds have been landscaped with shrubs, trees, and walks, and a bronze fireplace has been built where the original Lincoln fireplace was located. This site was originally marked with a cheap stone marker and the surrounding land was covered with weeds and rubbish."

With the cleanup complete, Frank presented the land to the state of Indiana as a gift. The deed to the property was given to Governor Harry G. Leslie in a formal ceremony at the statehouse in Indianapolis, June 6, 1929.

Other benefactions by the family cover a wide range of categories and interests. Among them were the Y.M.C.A. and Y.W.C.A. buildings and facilities in Muncie, the Masonic Temple and land for the American Legion of Muncie. The historic Gilbert residence, home of Muncie's pioneer family, and its surrounding property were purchased by E. Arthur Ball, past commander of the local American Legion and past state commander, and given to the Muncie post as its permanent headquarters. Also, extensive funds were given over the years to churches and various religious interests. The actual number of projects, both large and small, and the funds donated went often without notice or fanfare, according to the wishes of the family.

A touching and enduring memorial to the many philanthropies of the Ball family is the monumental sculpture standing on the Ball State University campus.

A move to recognize the many contributions to the community made by the Ball family was first begun in 1927 when civic leaders decided that some expression of gratitude would be appropriate. They approached the world-renowned American sculptor Daniel Chester French, best known for his sculpture of Abraham Lincoln in the Lincoln Memorial in Washington, D. C. By 1930 he completed the seven-foot bronze statue of a "heroic figure robed and winged, bearing in one hand a half-opened casket of jewels, her right hand outstretched in a gesture of bestowal." French named the piece *Beneficence*. It was destined to be his last commissioned work of art and to stand with five classic Doric columns as a symbolic backdrop.

French did not live to see his statue installed and dedicated at its eventual site. For seven years the statue was kept in storage during the darkest days of the Great Depression. Finally, on May 18, 1937, the Muncie Chamber of Commerce, headed by banker Robert H. Myers, launched a campaign to finish the tribute. It raised $31,965 from 11,306 persons. The statue was dedicated September 26, 1937, and the ceremony was broadcast across the nation. Although three of the original five brothers were dead by this time, a plaque on the pedestal below the statue reads, "A civic testimonial to the beneficence of the Ball brothers and their families."

In his address given at the dedication of the statue, Glenn Frank, then president of the University of Wisconsin, made the following remarks recorded in Frank's Memoirs:

"There is something unique about this ceremony. It is a little off the beaten track of the typical American reaction. We are, let us be honest with ourselves, a notoriously ungrateful people. We are, although we refuse to admit it, a notoriously inconsistent people. We worship, sometimes with absurd excess of enthusiasm, at the shrine of competence, efficiency, and achievement and then cock a suspicious eye at any of our fellows who reach positions of power through their manifest competence and efficiency. We spend half our time crying for great leadership and the other half crucifying great leaders when we are lucky enough to have them come our way. . . .

"But today Muncie breaks with this American habit of inconsistency and ingratitude in its common tribute to the Ball family. Here is a family that has displayed competence, efficiency, and achievement; and Muncie, instead of cocking a suspicious eye at the men who have been the bearers of this family tradition, has, through a medium financed by popular subscription, asked to say through a sculptor's art that Muncie is grateful. Here is a family that has given leadership in the widely varied fields of art, education, health and religion; and Muncie, instead of thinking this leadership has taken something from the community, insists upon going on record that this family has given, and given richly, to make Muncie a more congenial home for the human spirit.

"All this is a tribute alike to the city of Muncie and to the Ball family. The Balls have acted more as the social trustees than as the personal owners of the fruits of their competence. And Muncie has had the rare good sense to know good citizens when it sees them."

Today Ball State University students affectionately refer to the statue as "Benny." There is a tradition on campus that coeds best be kissed below her uplifted wings and after midnight.

Part Four

Trusts, Trains and Tyranny

The 1930s and 1940s were years of expansion and continued
prosperity of Ball Brothers. Fruit jar sales remained
important to the company and when the U.S. government moved in
to charge that abuse of patents can lead to monopolies, an eight-
year court battle resulted which involved most of the major glass
producers. With three of the original five ball brothers dead
by this time, Frank continued to lead the company during these last
years of prime fruit jar production. George, meanwhile, busied
himself in a financial adventure which had him running a large bloc
of the nation's railroads. Then the war came. Frank would die before
it was over. And then a new generation of company leadership
would emerge to face the mounting problems
in a new world of rapidly advancing technology.

At left is Frank C. Ball shown as he appeared in 1938 before a federal committee investigating trade monopolies. The year after this hearing, the fateful lawsuit was filed against Ball Brothers and other glass industry giants, U.S. vs. Hartford Empire *et al.* F. C.'s son, Frank Elliott (middle), the "crown prince" of the family and one-time heir apparent to company leadership, is shown in the garb of his beloved Masonic Order. He died at the age of 33 in an airplane crash near Findlay, Ohio. His mother, Mrs. Elizabeth Brady Ball, is shown at right around the time of F.C.'s death. Ever a devoted mother and wife, she was also active in civic affairs.

In his later years, George A. Ball became a newsmaker. *Muncie Star* columnist and amateur ornithologist Dick Greene remembers the casual relationship G. A. had with newsmen: "He'd see an unusual bird and just call to ask me something about it." G. A. is shown here giving what proved to be the last interview before his death in 1955. Those who knew G. A's penchant for collecting pencil stubs, old calendars, and time-hardened erasers, will smile affectionately at a telltale bit in this picture—a hole in the bottom of his shoe, just a step away from the difficult decision to buy a new pair.

A letter from Herbert Hoover shortly after the election loss in 1932 to Franklin Roosevelt bespeaks George's interest in politics. G. A. was a Republican National Committeeman between 1932 and 1937 and was active in the campaigns of other such Republican presidential hopefuls as Wendell Willkie and Alf Landon. Another historic letter survives in the Ball company archives, a personable note to G. A. from Theodore Roosevelt, post marked from Oyster Bay, New York.

Besides knowing well the political giants of his day, G. A. rubbed elbows with some of the most notable stars of the silver screen. Here George (far right) is shown with (left to right) film stars Robert Montgomery and Rosiland Russell and "movie czar" and Republican National Chairman Will Hays. The scene is unknown, but probably this was a joint public appearance, a crossing of politics and Hollywood, as the more thoughtful stars became socially conscious. (Montgomery is not in a bathrobe: the buttonless, wraparound overcoat was a top fashion of the day.)

THE WHITE HOUSE
WASHINGTON

November 30, 1932

My dear Mr. Ball:

I wish to convey to you my appreciation of your effective leadership in the campaign, - a campaign against great odds. The election result should, I believe, be taken not as a discouragement to the Republican Party but rather as a challenge to continued zealous and aggressive work in behalf of its sound and enduring principles. I wish to thank you personally for your able and unselfish work in behalf of the Party and myself.

Yours faithfully,

Herbert Hoover

Honorable George A. Ball,
Republican National Committeeman,
P. O. Box 545,
Muncie, Indiana.

For a short time during the 1930s, a member of the Ball family became known nationally for something other than fruit jars. It was a saga that began when the Van Sweringen brothers of Cleveland drove to Muncie one winter morning to seek financial help from George A. Ball to save their faltering railroad "empire." "If you don't help, we're stopped dead," they pleaded. G. A. said later, "If those properties were sold in the open market, . . .they might fall into the hands of crooks and amateurs and chiselers. . . .The ruin might embrace the nation itself." He played "Santa Claus," as one newspaper described him, and helped out. Then came a strange turn of events. The brothers died within months of each other, leaving G. A. the owner of their tangled "empire." "Mr. Ball Takes the Trains," *The Saturday Evening Post* put in headlines. Indeed he did, but to Washington to tell his strange story to inquisitive senators there. He told them candidly he knew little about the properties he now controlled and impressed them with his quiet humor and honest sagacity.

Point of discussion: Max Lowenthal, committee counsel; G. A.; Senator Wheeler, chairman; and Senator Wallace White of Maine. Committee members were fascinated by G. A. who "gives away $100,000 at a time and carries his money in a small brown purse."

Senator Burton K. Wheeler, chairman of the investigating committee, and G. A. study a chart of the intricate Van Sweringen interests. There were 246 corporations listed, but many were missing and could not be traced. The "empire" included an estimated 28,000 miles of rail lines.

The Ball family made two philanthropies of historic significance relating to U. S. presidents. Above, F. C. Ball presents to Indiana Governor Leslie the deed to the property on which stands the Hoosier cabin where Abraham Lincoln spent his most formative years. Below is the house George Washington bought for his mother in 1772. The residence of Mrs. Washington, an ancestor of the Ball family, was restored with Ball help. These two contributions are representative of the family's generosity in many areas, among them education, religion, and health.

In recognition of Ball philanthropies, Muncie citizens planned this tribute, *Beneficence*, the last commissioned work by noted sculptor Daniel Chester French. The sculpture was completed in 1930, but the Great Depression intervened and it was seven years before its eventual dedication in a ceremony which was broadcast on national radio. The statue stands today on the campus of Ball State University and is known to thousands of campus residents and visitors as a symbol of the university and family spirit.

From the pages of the *Ball Line* come the war years at Ball Brothers. At left Paul Miller, a stockman in the corrugated box department, proudly sends his six sons off to war. The small print below his picture reads, "His only regret is that he himself is unable to get in there and help rid the world of dictatorship." At right is Seaman First Class Dorothy Ruble, home on leave, as she drops in to visit her old friends at the company. Dorothy was employed as a sorter in No. 1 Packing Room before enlisting in the WAVES.

"Jake," a thoroughbred collie, at right, served his country for two and a half years in the "Army of Dogs for Defense." He was enlisted in the armed forces in 1942 and received his basic training at Fort Robinson, Nebraska. A sentry dog, Jake was the property of John H. Schench, whose father, Holmes, worked in the main machine shop at Ball Brothers.

During the war years future company president John W. Fisher assisted then-president Frank C. Ball (right) in operating a company largely dependent on personnel unqualified for military duty and by relief workers brought in to fill the ranks. Abroad, Major Ed Ball served in various military capacities, among them in the Fifth Army under the command of General Mark Clark. Here Ed (below, third from left) walks literally in the footsteps of the famous commander as they review a French Air Force detachment at Marrakesh Airport, Morocco, prior to the invasion of Sicily.

A number of working women have played a large part in the company's story of success. Here, operating the switchboard during the post-war period, is Mary Cates (left) who served as receptionist and, later, secretary to Edmund F. Ball. At right is Maxine Williams, who was the cheerful voice of "Ball Brothers" and later "Ball Corporation," handling hundreds of incoming calls daily.

Ever-loyal Gertrude Barrett began at Ball Brothers as an office girl at the age of 16, riding the streetcar to and from work each day. Prior to her death in 1964, she had served as confidante, secretary to George A. Ball, and family and company historian. It was her meticulous notes and voluminous scrapbooks that provided important bits of remembrances invaluable to this history.

It was said of "Aunt Mary" Lincoln's birth that she brought sunshine into her parents' home. The same was said of her at Ball Brothers. "Aunt Mary" came to Ball Brothers in 1906 as a cashier. In the earliest years it was a familiar sight to see her driving to work in her "shining" electric car. Years later she became a legend at the company and her birthday became an important event each year. On one occasion, one writer for the *Ball Line* was driven to quote Robert Burns to fully describe the universal affection for her at the firm. He wrote: "And we will love thee still, my dear/Till all the seas gone dry."

Chapter 11

THE WAR YEARS

After the Japanese bombed Pearl Harbor on December 7, 1941, the United States as quickly as possible shifted into a wartime economy.

Ball Brothers was in the thick of it. On General Lewis B. Hershey's list of activities essential to the war effort, drawn up in the summer of 1942, were four industries related to Ball Brothers: the manufacture of fruit jars, glass containers, zinc shells for batteries, and various rubber products. This meant that Ball Brothers was eligible for basic materials which were generally now in short supply, but would be issued to Ball plants by the War Production Board.

The excitement of the war and the demands made upon the entire company for maximum output obscured all other concerns as the company buckled down to do its best.

Colonel Alvin M. Owsley spelled it out at a Muncie meeting of company superintendents and foremen: "Our zinc division is furnishing millions of battery shells a year and the company will soon spend thousands of dollars to increase this production to hundreds of millions."

Colonel Owsley, who left his law practice in his beloved state of Texas at the request of his father-in-law to assist in the management of the company during this difficult period, covered other areas. He said then: "This company has been compelled to desist in the manufacture of our world-famous one-piece, porcelain-lined cap. It is necessary to do this because our fighting forces need zinc to make shell casings. In the early drive of the Germans into Russia, our Russian ally lost her zinc mines, and under the terms of the lend-lease program we must make the loss good. Ball Brothers, being a major user of zinc, will be hard hit, but has cheerfully given up zinc. Mr. F. C. Ball, president of the company, through his representatives, offered this sacrifice to the government before it was demanded." The company also cooperated with officials in dealing with the rubber shortage.

The Ball machine shops, meanwhile, were in the thick of the war effort. They were making parts for 30- and 50-caliber shells for Waterbury-Farrel Foundry and Machine Company in Waterbury, Connecticut. They made parts for the Graham-Paige automobile company in Detroit and for Wright Aeronautical Corporation in Dayton, Ohio. Ball Brothers jigs and fixtures were used in the manufacture of diesel engines for landing barges. It made parts for the M-3 tanks turned out by the Pullman Standard Car Company. One of the most effective Ball Brothers war projects was to supply side notch grinder and counter bore machines from designs created in the Muncie shops. The end product, produced at Perfect Circle Company at Hagerstown, Indiana, a firm which originally manufactured piston rings for automobiles, was the aircraft piston ring, which was high on the War Department's demand list.

To help keep up with demands, the company made two important acquisitions during the war. In 1942 it purchased the Aridor Company of Chicago, Illinois, a manufacturer of closures for packersware. In 1943 the company also purchased the American Zinc Products Company, a subsidiary of DuPont located at Greencastle, Indiana. This acquisition added substantial zinc rolling capacity for battery can production which was increased substantially.

During the same period, the company was losing some of its strength at the top. Three of the founding brothers had already passed on. Now, in the midst of the crucial war years, in 1943, Frank C. Ball, president of the firm since its first day in business 63 years before, succumbed at the age of 85, a strong personality even to the end. There are stories still told of F. C. directing company operations from what would prove to be his deathbed. After his passing, G. A., the youngest and now the only survivor of the five Ball brothers, was to head the firm at 80 years of age. In spite of his own advanced years, G. A. had always been active in all the affairs of the company, rarely missing a day at the office, a habit he was to follow even after he passed his 92nd birthday. Characteristically, G. A. declared with an absolutely straight face that his reputation as a hard worker had been vastly overstated. "I have always worked hard," said Mr. Ball. "I still do. But I never worked more than 24 hours in a day, and I don't now."

Full-time work was demanded of him to solve the many complications the war conflict was bringing on. Besides such key materials as glass, zinc, and rubber, the war needed some of his essential plant personnel. The company made substantial contributions in this area. From the Ball

"family" of some 3,000 employees, hundreds of trained men and women went into the armed forces, by draft and enlistment. The records of the company during those war years are filled with pride, joy, and sadness. The *Ball Line,* the employee newspaper originated by John Fisher, turned literally into a photo gallery of young men and women heading out into service. The honor roll is too lengthy to include here, but one's thought's linger over Frank Collins, a veteran of the Number 2 Packing Room in the Muncie plant, a widower for many years, who had raised five sons alone in his house on South Madison Street. Now he was even more alone. Every son was in uniform. Paul Miller, corrugated-stockman in the box department, had *six* sons in uniform. His chief regret was that he couldn't get in there with them, being past military age. There are *Ball Line* photographs of those returning to visit their old buddies in the plant, and there are photographs of those who would never return again.

The war took some immediate members of the Ball family as well. Vice president of the company E. Arthur Ball had already served his country in World War I as part of the famous 42nd "Rainbow" Division fighting in France. Now, as Captain Ball, he was back in uniform again. He entered service in 1942, was promoted shortly to major, and saw service in North Africa. He was then transferred to northern Italy as a military government officer, went on to England six months later, just in time to land in Normandy with the 80th Infantry Division. Then it was Luxembourg, Austria, and Germany. Just before entering Germany, he attained the rank of lieutenant colonel.

Edmund F. Ball was 36 in 1941. He had entered the family business after earning his Ph.B. at Yale in 1928, and then spent the next eight years at various production assignments in the glass factories of Ball Brothers' main plant in Muncie. In 1936 he was named manager in charge of the six glass plants which Ball Brothers was operating in Indiana, Oklahoma, Texas, Illinois and West Virginia. He was a young man whose education included attendance at the Asheville School for Boys in North Carolina, and Indiana's Wabash College, in addition to his Yale experience. In spite of the fact that he threw himself into all of the most technical operations of Ball Brothers and learned them right down to the nuts and bolts, his was a studious disposition, inclined to literary, philosophical, and social pursuits, very much in the manner of his own father and his four uncles. But those hard-nosed years in the Ball plants were going to stand him, and the company, in good stead in the future.

Prior to the war, Ed was serving as assistant secretary and a director of the company as well as being responsible for the operation of its various glass plants. Some years previously, he had become an officer in the Army Reserve Corps, enjoyed its various activities and associations, and was commissioned a first lieutenant in the Infantry Reserve. But he had not been active in the organization for a decade. Consequently, he was astonished when he received by registered mail, June 24, 1941, a communication from the War Department ordering him to active duty—six months before Pearl Harbor. The term of active duty was then thought to be for one year. "Had my mind been more solidly fixed on what I wanted to make my career, I might have asked for deferment," Ed said years later. "I talked it over briefly with Uncle Frank and Uncle George, my mother, my wife Isabel, and other members of the family. I convinced them and myself that it would be a good experience, that I would only be gone for a year, would be close to home, and that I should get the obligation I had assumed as a reserve officer over as promptly as possible. Perhaps in the long run this was a fortunate decision."

John Fisher, Ed's brother-in-law, was midway in his studies at Harvard Graduate School of Business Administration, but was prevailed upon to cast his lot with the company. Ed did all he could to fill John in on glass container production and operations in the little time left before reporting for active duty. He also made arrangements for John to cover for him on other outside interests, such as his farm, the Muncie airport and Muncie Aviation Corporation.

Ed would be in uniform for four years and see service of many kinds in many places. He first attended the Army Industrial College in Washington, D.C., which he described as "a combined Harvard Business School, Command and General Staff, and Army *Emily Post*." After completing that course, he served on the General Staff before going overseas in June, 1942. His experiences as a staff and air-ground support officer were to bring him into contacts with enemy infiltrating patrols, and shore battery blasts directed against the division to which he was attached, all of which he shared with the humblest of fighting men. He was also in a remarkable position to see World War II, the greatest mass enterprise ever attempted by man. It may well have been the finest training ground a young American businessman could possibly have undergone.

It would be needed at home.

Problems were beginning to surface in Muncie. While he was serving as Air Support Liaison Officer with Sixth Corps Headquarters, part of the Fifth Army, located just outside the little town of Prata in Italy in the winter of 1943, the first inkling of this reached Ed. He recalls, "It happened to be November 11th, Armistice Day, when I received a very discouraging letter from Uncle George telling me of all the troubles and problems the company was having and urging me to try to find some way to get a furlough and come back to discuss business matters. By coincidence, General Clark, whose father-in-law had been a long-time salesman for Ball Brothers and knew the family as well as me, happened to visit Sixth Corps Headquarters a few days afterwards and I had an opportunity to briefly share the letter and its contents with him. He was most sympathetic but doubted very much whether he or anyone else could arrange for a furlough for me back to the States. He did, however, suggest the possibility that I might be eligible for rotation before too long.

"It was not until the end of December that I made the definite decision to try to be rotated at the first possible opportunity so that I could get back to the United States and be more available for business conferences from time to time and, of course, more importantly, to be back with my family again. It was not until after another D-Day landing at Anzio and about eleven weeks on the beachhead that I received my orders that I was being rotated and would shortly return to the United States. Actually I didn't get back to the United States until toward the end of April and was eventually stationed at Ford Motor Company River Rouge Plant, Detroit, which of course was much nearer home but still pretty far away for all practical purposes."

On January 2, 1945, Ed received a telephone call from John Fisher informing him that their continuing enemy, fire, had hit the Number 1 Glass Plant in Muncie. The building had burned all night.

Ed got a day's leave of absence from his boss, Colonel Hart, obtained an airplane from his operations officer, and flew to Muncie. It was a bitterly cold day when John met him at the airport and drove him to the plant. A terrible-looking situation waited them. The packing room was destroyed and, because of the heat, the supporting columns had collapsed so that the roof over the blowing room was actually resting on the tops of two furnaces. The Austin Construction Company was called in and a remarkable job was done in shoring up the structure, so that very little production time was actually lost. It was because of the skills the Austin company demonstrated in handling this problem that they were later given the contract to build the new glass plant in El Monte, California, in 1947. The loss of $500,000 fortunately was covered by insurance.

Finally, on February 9, 1945, Ed's resignation from the Army and Army Reserves was accepted. He was separated from active duty at Camp Atterbury, Indiana, and returned to civilian life.

As Ed was to write later in his memoirs, *Staff Officer*

"My job then was in Muncie, and I got back to it as soon as I could. When I learned firsthand the incredibly tough job our people had been doing in keeping our plants operating as shorthanded as they were, bedeviled by regulations, shortages, restrictions, rationing, demands for this and demands for that, I almost felt like a shirker, for certainly the jobs done at home were equally as important and difficult (although not quite as dangerous) as the jobs done by those in active service."

Chapter 12

CRISIS AND RENEWAL

Ed Ball is proud to say that Ball Brothers has, throughout its history, been "beset in Muncie by the same things—fires, panics, depressions and other adversities—that have ruined many a bigger business. But it survived—sometimes on sheer grit—and the family has lived to see the company take its place among leading industrial enterprises."

But when Ed returned from the war, he encountered a family-owned firm that had been ravaged by more than just a fire or affected by financial decline. Something that had been happening over a long period of time suddenly became apparent to everyone, a situation no doubt accelerated by the devastating effects of the world conflict. *BusinessWeek*, reviewing the company's post-war troubles, wrote in February, 1952:

"As the years went by, Ball Brothers began lagging behind more aggressive competition, like many other family businesses. By 1946, all but one of the founders was dead, the company was aging fast. At that point, Edmund F. Ball, son of one of the founders, returned from the war. Ball took a long look at the company, and didn't like what he saw."

What Ed saw when he returned was in part a glass plant that was in particularly bad shape. One veteran employee remembers how the Muncie facility and others were just about consumed by the unremitting demands of the war period. So were the employees. "The younger people were taken, the older people became exhausted, and eventually we had got down to what was left," Ed remarked.

For example, on a management level, there was no such thing as inventory control. It was a small matter, but one employee noticed that coins mailed in for the purchase of the popular Ball *Blue Books* were not deposited promptly at the bank. They were just put in a sack and stored in a corner. "When checks came in and were opened, nobody put any restricted endorsement on them, and they might lie around for two or three days before anybody would bother to go down to the bank with them," one Ball executive remembers. "Accountants sat around with celluloid eyeshades and garters on their shirtsleeves. It was a wholesale grocery office out of 1910."

These are long-forgotten details, but they were indicative of larger management problems which were beginning to hurt.

Ed summed it up with a rueful phrase: "It was a fine old business that, frankly, had begun to go to seed." Another company executive looking back remarked more harshly, "The company was dying."

These were stark observations, but they stated plainly the company's most critical period of crisis. According to one observer, Ball Brothers was asleep at the switch during the war, and when things went back to a buyer's market in the post-war period, Ball Brothers had not changed its old posture or method of doing business. It was also burdened by a huge inventory of glass containers which no one wanted in the new market. The opinion was offered that Ball Brothers, for better or worse, had been trapped into being a marginal supplier in an expanding commercial market; that in fact (and most damaging of all) customers were buying from Ball only if containers were difficult to get elsewhere.

Years later a Ball executive commented, "By the late 1940s all these things had snowballed into a situation where they could say, 'OK—Now what are we going to do?' "

The answer was going to be up to Ed. But before he found it, he had to survive the disillusionment of coming back home from the service to a business which his father and uncles had developed into a monument, the foundations of which were now beginning to show cracks of wear and tear and obsolescence. It was a painful reintroduction.

"When my terminal leave was ended and I returned to Muncie, I was almost completely overwhelmed by the task that seemed to lie ahead," Ed said. "Here was a family-owned company with a fine reputation and a history of making money. But the plants and equipment were run down, badly needed replacement, major repairs and modernization. With a few notable exceptions, the organization was old and tired."

There was some momentum left at the close of the war. It will be noted that while 1946 and 1947 were not bad years, the trend was established and the certainty of many serious problems ahead was clearly evidenced by drastic decline in profits to a virtual break-even point in 1948 and a loss of almost a million dollars in 1949.

The figures given here tell the story of the nine very difficult crisis and reconstruction years between 1946 and 1954.

Year	Net Sales	Profits Before Taxes
1946	$32,562,797	$1,167,975
1947	32,187,844	1,223,159
1948	35,277,405	333,534
1949	30,752,591	(913,375)
1950	38,497,086	1,476,851
1951	42,698,610	2,160,096
1952	37,736,955	(739,471)
1953	46,719,587	2,087,218
1954	44,197,870	163,036

"In the years to come I often thought it might have been easier to have started an entirely new company than to restore an old one," Ed commented. "But it still was a going concern with major assets, a fine reputation and a willing organization."

Shortly after his homecoming, Ed was elected executive vice president at Ball Brothers. Within a couple of years an effort was made at expanding the company's glassmaking capacity when a market survey showed the possibility of additional sales for commerical ware on the west coast. Here seemed an excellent opportunity to test the firm's mettle and see if it was really up to the challenge of total rehabilitation. In 1947 a new glass plant was finished at El Monte, California. John Fisher had overall responsibility for the construction of this plant, one of Ball's finest. It contained the first earthquake-proof glass furnace on the west coast, and other innovative devices and systems such as an automatic batch-weighing system, electric heat boosters on its furnaces and electronic controls. Always a leader in the forefront of modern glass technology, it was among the first to experiment with and adopt sophisticated emissions control processes in compliance with stricter environmental requirements.

Encouraged by the El Monte venture, but discouraged by the overall financial picture of the company, Ed faced some difficult decisions when in 1948 he became the third president of Ball Brothers. One of his first moves was to present to the board of directors the alternative courses which were facing the company for the future. These were the choices as Ed saw them:

1. The company could be liquidated and the proceeds divided and distributed to the individual stockholders on a per share basis to do with as they pleased.

2. The company could be sold as a going concern which would also provide funds to the individuals, much as in the first alternative.

3. A merger could be sought.

4. Management could continue to operate the company, do its best to restore it to a successful operation, and preserve the family name and tradition.

Ed left no doubt as to where he stood. He said, "I made it clear that I would have no part in the liquidation of the company, that I would not like to see the company sold but if asked to do so, would do my best with whichever of the other three alternatives might be determined. I stated that although I realized it would be a long, hard struggle and results might be disappointing, I favored the fourth alternative and outlined my reasons. Some, perhaps for personal reasons, might have wished to have funds available to support their own special interests, but nevertheless the #1 alternative was not seriously considered. It was ultimately decided, unanimously, to select alternative #4 and do the best we could to make a go of it on our own."

Commenting on the situation later, in *Nation's*

Business magazine, John Fisher said:

"The business could have been sold or liquidated without financial hardship on the family. But turning things around would require hard work and tough decisions, some of which would have an adverse effect on Muncie, a community to which the Ball family had close business and personal ties. In deciding to meet the problems head on, Ed Ball and other members of his family acted in the competitive spirit of the founders . . . and reflected their determination to manage change rather than be overcome by it."

It soon became clear that the long road ahead would require stamina and fortitude. Ed was up against some formidable obstacles. "After the decision was made, incredibly difficult problems became painfully apparent," he recalled. "Although wishing to cooperate, people were set in their ways. It was difficult to bring about changes. The old ways had always been successful in the past and therefore were good enough for now and the future. Why change?

"The company had a very strong balance sheet with a ratio of current assets to current liabilities of seven to one. It had the reputation for always making money. Why was there any concern about profits or cash?

"The trouble was that most of those current assets weren't current at all. Huge sums were tied up in inventory, which in those days only had a turnover of about one to one-and-a-half times per year. Receivables were high and, while mostly good, were slow paying.

"It was almost impossible to get the point across to those involved that we were virtually out of cash. There were periods when it was nip and tuck as to whether we would have enough cash to make the payroll, an obvious fact which many of those involved refused to believe. When business began to fall off after the war, plants remained running, building up huge inventories. Operations were only closed when there was no storage space left to put the product. Customers were accustomed to long and favorable terms and, probably facing cash problems themselves, took advantage of the long terms before making a payment."

It was a cruel dilemma. There is eyewitness evidence to the fact that George A. Ball himself was visibly shaken by the inadequate financial resources the company had available at the time. The organizational crisis might be overcome. Ball might produce new products for new markets, but first the money had to be there to make it all come about. However, conventional efforts to raise capital were unrewarding.

A good many opinions have been expressed about this crucial juncture in the Ball Brothers' saga. There are many records and reports pertaining to what happened, proposals of various kinds, examinations of collateral, financial needs, and concern of various banking groups as to adequate security should they advance funds to the corporation.

John Collett, then a director of the company and its financial adviser during this difficult period, agreed with G. A. that, after all of the company's liquid assets, inventories, and securities had been liquidated, for all practical purposes then—"We're broke."

Except for one thing. The Ball family's reputation for entrepreneurial ability, integrity, indefatigability, technological innovation, fair dealings, honesty, and—pardon the expression—sheer guts, won the day.

John Collett, whose firm of Collett & Company, Inc., of Indianapolis made the arrangements for a loan, often tells the story of the financial deal made with the John Hancock Life Insurance Company of Boston, which salvaged the situation. "The president of one of the biggest insurance companies in America knew Ball Brothers' reputation," Collett recalls. "Things were not particularly good over the country at that time, and Ball's earning figures weren't very good either. But we asked for a loan of $6 million. I'll never forget it. I was driving from Muncie to Indianapolis with an official of the insurance company in my car, and just as we passed a certain big tree along the side of the road, he said to me: 'John, the company is going to approve a $6 million loan at 3 percent interest for 15 years.' "

Collett says: "I never drive past the tree now without tipping my hat to it."

But there was still another problem to be faced. Almost as an afterthought, the John Hancock representatives asked to review the company's insurance. It carried virtually none. Frantically, to finalize the loan, policies amounting to $25 million of coverage were arranged for and purchased. Some of the first funds received from John Hancock were to pay the premiums. The Hancock loan ultimately went through on May 29, 1948.

The money would certainly help. But it could not do everything, as Ed Ball was all too well aware. Now he had to face the hard fact that Ball Brothers would have to abandon its traditional family management and bring in outside personnel.

Ed recalls, "I had determined fairly promptly that we didn't have the capabilities within our organization to objectively analyze our problems and determine best courses of action that should be taken. After considerable thought I brought in, for better or worse, a management consulting firm by the name of Robert Heller & Associates of Cleveland. They made an extensive and expensive study with many recommendations. There was a considerable amount of criticism of the so-called Heller Report, much of it justified, but it did serve the purpose of breaking through the shell of indifference that permeated the organization, bringing to light our many weaknesses and jarring the organization out of its complacency."

The report by Heller held some bitter pills to swallow. Four basic problems were isolated:

1. The company's capacity for production was ahead of its marketing.

2. The volume produced was high cost with too-low gross profit.

3. Controls were lacking. Controls over production, costs, personnel, inventory and sales. Management needed controls to administer affairs of business.

4. Organization was inadequate, particularly as regards planning. Planning was not organized in the various divisions of the company and the vehicle for overall planning did not exist.

There is not much doubt that it was precisely at this time that the Ball Brothers company faced the music, courageously accepted the challenges contained in the Heller Report, and went on to do something about them. It is clearly one of the most controversial periods in the entire history of the company. Understandably, many of those who were with the firm were not in agreement as to how it was to be done. But the young army-officer-suddenly-turned-corporate-executive had made his decision. The company's future was again in the making.

Ed has called the year 1949 "probably the most difficult and the most discouraging of my entire life." The company was involved in very serious labor negotiations in March of that year with the Federation of Glass, Ceramic and Silica Sand Workers, CIO, which had won the employees' representation rights in the Muncie plant.

Alec Bracken, George Myers and John Fisher were the major negotiators for the company and Joe Frosch, national president of the union, was the principal negotiator on the union side. Deadline for the strike had been set at midnight and Ed was sitting with the negotiators doing everything he could to avoid a shutdown.

"I knew that our situation was so precarious that a strike of any seriousness or length might very well be the deathblow of the company," Ed commented years later. "I was terribly concerned as were we all."

Ed observed that Joe Frosch seemed to appreciate the seriousness of the situation and also wished to avoid a strike.

"He was as rough with his people when they got out of line as he was with us," Ed said, "but there was never any doubt that he controlled his union and that he was a man of his word."

As the midnight deadline approached, Frosch turned to Ed and asked for his opinions. Although no exact record of his remarks has survived, Ed recalls he appealed to his employees to stay on the job and spoke at length on the seriousness of the situation and what a strike might do to the company and to them, the employees. Finally, just before the deadline, Frosch banged on the table between them with his huge fist and said, "That's all I want to hear! You've got a contract!"

"That could well have been one of the most fateful decisions in the history of the company," Ed said years later. "Had a long strike resulted, it might have been a fatal blow, the situation that year was so precarious."

After the strike was settled, Ed was physically, mentally and emotionally drained. He and his wife Isabel then decided on a brief vacation and left the next day for Fort Myers, Florida, to join long-time friends Ray and Katherine Applegate for a few days of rest and fishing. Instead of rest and pleasure, however, the episode turned to tragedy. The fishing boat which they had chartered exploded. Katherine died in a very few minutes and Isabel, fatally injured, survived only until the next day. Ray and Ed were seriously injured, although both eventually recovered.

The period of recovery turned out to be a difficult one for Ed. In addition to his physical and emotional injuries, he was now left with three young children to raise and a commitment to a business that was losing money and needed heroic efforts, almost miracles, to occur if it was to survive. G. A., Alec Bracken, John Fisher, George Myers, Fred Petty, and others rallied around him to keep things going.

But soon a second tragedy occurred. Fred Petty suffered a fatal heart attack while at his desk on September 30, 1949. Although the company nurse and others did all they could to revive him, he died before the doctors could arrive.

Fred Petty's death, according to Ed, was "the last blow."

"It was more than ever apparent to me that it was absolutely necessary to bring in additional help to strengthen our organization," he said. "It was painfully obvious that too few people were trying to do too much and that the company would never grow—perhaps not even survive—if its organization was not strengthened."

There were even problems under the surface which loomed fully as large as facts and figures and strikes. The changes which were so desperately needed on all sides were blocked, often without definite intent, by complacency and inertia on the part of those both inside and outside of the company who had come to some firm conclusions difficult to dislodge.

Ed Ball clearly recognized what harm these influences had upon his decision to go forward with the times—but doing something about them was another problem.

It was, almost tragically, the reverse side of the coin which had made Ball Brothers such a prosperous and respected company.

What might well have been considered—and was at the time—a major strength turned out to complicate and make the problem more difficult: "the family image."

As the public saw it, Ball Brothers Company appeared to be a firm operated successfully by its family owners. There was ample evidence to support this belief. All members of the family were thought to be affluent. The family had made the decision to continue to operate the company and not liquidate, sell or merge it. Furthermore, it appeared that it was to continue to be operated under the leadership of the family as evidenced by younger members being placed in top management positions. As far as the public was concerned, the company was to remain for all the foreseeable future, closely held, owned, operated and managed by the family.

This family image created many serious problems detrimental to the implementation of plans, and the execution of what must have appeared to be arbitrary and drastic changes necessary to turn the business around.

Such a simple and basic need as to operate the company profitably was looked upon with skepticism by many. Why were profits being discussed when it was a well-known fact that the company had always made a lot of money and the family was already wealthy? Many people found campaigns to reduce costs difficult to appreciate in the face of the family's record of benefactions. Some even interpreted these moves as simply an effort on the family's part to acquire even more personal wealth.

Employees had difficulty comprehending the problem. Ed remembers: "I was shocked when discussing the company's problems with a key employee whom we considered might be qualified for more important positions in the organization. I was emphasizing the importance of profits and he responded, 'I've never had anything to do with profits and never will!' He was responsible for expenditures of large sums of corporate funds, deliveries of supplies, materials, services on precise schedules, and accurate specifications. And yet he didn't feel he had any responsibility for producing profits. The impression was all too typical throughout the organization."

Wherever Ed sought capable persons to help strengthen the organization, he was asked the same question: Could an outsider really find a place in the organization? It was difficult to interest individuals who might otherwise have been qualified for advanced management positions.

Nothing could have seemed more complex or sensitive. It was to tax all of Ed Ball's ingenuity and diplomacy.

Furthermore, the problem, ironically, was occurring at a time when some members of the family wished to pursue other interests. William H. Ball moved to Indianapolis and chose to devote his time to personal pursuits. Alvin Owsley reached the decision to return to Dallas and his beloved

Texas from which he had come at the urgent request of his father-in-law. E. Arthur Ball, who had just returned from military duty in World War II, became interested in a small glass fiber manufacturing company by the name of Friedrich and Dimmock, located in Millville, New Jersey. In January, 1947, G. A. and Ed made arrangements for Arthur to take over this assignment and establish his headquarters in Millville. It was there that he died suddenly on April 3, 1947. He was found in his car at the airport where he kept his small plane. Not too long afterwards, the Friedrich and Dimmock operation was sold.

When on January 15, 1948, Ed became president of the company, G. A. went into semi-retirement as chairman of the board. According to Ed, "He remained for the rest of his life in a valued advisory position. He was always helpful, but I don't think he ever felt sure that it would be possible to do all that needed to be done and all the problems solved necessary to make the company really successful."

Ultimately, it was decided to find a firm which was successful in the field of recruiting management talent. One of the most prominent so-called "head-hunter" firms was Jack Handy and Associates of New York. It was to this company Ed turned for the outside help that was needed so desperately. He made clear that he wanted an executive vice president with experience in modern management, someone who could put together an organization which would best utilize present employees as well as include any further outside help considered necessary. The man was to help lead an overhaul job that Ed envisioned would take about five years to accomplish.

Handy presented Ed with several candidates for the job, but two of them that Ed particularly liked declined to accept the position when it was offered.

"I am sure they were disturbed by the company's apparent prospects and really wondered if it could survive," Ed recalls. "They were concerned first of all that it was a family-held company and that an outsider would be subject to the whims and wills of the family. Furthermore, the company itself must have looked about as attractive as a buggy whip manufacturer with its principal products being home canning supplies (which appeared to be a declining business going out of style), a zinc operation which was about as exciting and appetizing a challenge as a mud pie and a rubber operation in a field that was known to be highly competitive and a cat-and-dog business at best."

Finally, the name of Duncan C. Menzies, a man who had come to this country as a lone immigrant boy and had worked his way up through the ranks of industry to an executive position with Johnson & Johnson, came to Ed's attention. He had excellent credentials, was highly recommended by his previous employers and was looking for the real challenge which the Ball Brothers' situation seemed to present. After numerous discussions and meetings with key employees and the directors of the corporation, an employment agreement was reached.

Menzies was strong medicine for the ailing firm. He was a man of such complex personality and controversial theories that it is virtually impossible today to find any two estimates of his contribution to the firm that conform in any way. One thing seems sure: Menzies was a high-powered, hard-driving, self-confident enthusiast. Perhaps that kind of man, at that time in the company's years of uncertainty, was needed for the job.

Menzies was a tough operator. He could study financial statements and find the trouble spots. At the time he first joined the company in 1950, the firm was in the throes of improving its financial position through liquidation of inventories, collection of receivables and so forth. It was then Menzies looked around and made the remark often quoted today that "the company had money eating its head off."

Menzies was afforded a royal welcome as the new executive vice president and general manager. Ed Ball introduced him to the company with praise: "We went looking for this man. We knew we needed someone to help out with all the complex and complicated phases of running a modern business such as ours.

"We knew that those of us who were handling the top problems of the company couldn't possibly cover all the situations that needed our full attention. We wanted a capable, fair-dealing person thoroughly experienced in manufacturing, in employee relations, in sales, financing, industrial organization and one who knows how to use tools of modern management."

Menzies replied with equal graciousness. He publicly praised the Ball family for the fact that its name and products were familiar in every state of the Union and in many foreign countries. He took note of their community associations and generosity and quite fittingly and admiringly compared them to the Hershey family of Pennsylvania, the Rowentrees of England, and the Lever brothers, all closely identified with their home towns and for sharing their prosperity with their fellow citizens.

Menzies gave a short resume of his business philosophy in the *Ball Line*:

"*The primary objective of business is profit, for*

without profit there can be no security, opportunity or growth for either the employee or the company. It was that great labor leader Samuel Gompers who said that the greatest sin management could commit against its people was to run an unprofitable business.

"A modern business is like a three-legged stool representing labor, management and capital. Weaken or destroy one, and the stool collapses. Part of my job is to keep an even balance among all three to provide maximum stability and solidity.

"Without customers we have no business."

Menzies went to work to get those customers. He emerges with a controversial but memorable image. Short in stature but literally bursting with energy, he was given to shooting from the hip on certain decisions, but it was a matter of pride with him to get it over and done with fast, whatever the problem was, a characteristic which seemed to balance off fairly well between drawing associates closely to him or sending them scurrying for the exits. In a phrase, Menzies was not a man to be taken lightly—he was, in truth, a powerhouse of abrasive charm.

To be fair about it, there were a number of unhappy decisions to be made and some unpopular things to be done, and quickly. The members of the Ball family, who had grown up with the business, had always been just a few steps away from the shops even while in the executive offices. Being on a first-name basis with hundreds of employees was a company tradition. That made Menzies' job far from a pleasant one. As one veteran says of those days: "There was a lot of dead wood to be cut out. There were shins to be kicked." It was up to Menzies to do the cutting and the kicking. Even some of the most loyal old employ-

ees had their conceptions of job tenure handed on from their fathers before them. You stayed on a job until you were too old to carry on. They couldn't comprehend why a man willing to work, no matter what his age, shouldn't be allowed to go on as long as he could carry out his assignments. There were forced retirements now. There were pensions to be paid out of nonexistent pension funds.

Menzies put his head down and drove on. He was a realist. "Business must make a sound profit," he said. "Reserves must be created—research must be carried on. Adventurous programs developed and mistakes made which must be paid for later. Bad times must be anticipated. High taxes inevitably will be paid. New machines and equipment of necessity will be purchased; factories expanded and remodeled; new products developed and new sales plans launched. We must experiment with new ideas."

Menzies moved under the watchful eye of President Edmund F. Ball, but he moved. First he created a batch of new jobs in a reorganization of top management. He split the company into six divisions: rubber, metal, paper products, and three geographical divisions for the manufacture and sale of glass.

It is said that Menzies might take six months and wade through 250 interviews to find the right men for the jobs he created. The new manager had a theory, heretical to some of those in Muncie, that it was entirely unnecessary to know anything about the glass business in order to run a glass business. The men on his new team, derived from diverse backgrounds, put that theory to work.

Ralph C. Edgar, former personnel director for Alleghany Ludlum Steel Corporation, Pittsburgh, was named vice president in charge of industrial, personnel and public relations. At that time Ball Brothers had 3,500 employees in Muncie, a fair chunk of the city's 60,000 population.

Fred A. Schlosstein, formerly of St. Louis, resigned a 27-year-old association with Price Waterhouse & Company, a firm of national repute in the world of public accounting, to become a vice president and controller at Ball Brothers.

Robert W. Biggs, a former Pittsburgh steel man who had more recently been operations manager for National Electric Products Company, Pittsburgh, came on board with Ball Brothers as vice president and glass manufacturing manager. James L. Knipe, a Union Paper Bag & Paper Company executive, was brought in as vice president and general sales manager, the fifth outsider to get a key post at Ball Brothers. Both Biggs and Knipe served on the board of directors.

Another important man on the team, although brought to Ball Brothers prior to Menzies' arrival, was G. Fred Rieman, who had resigned a vice presidency with Anchor Hocking Glass in 1939 to join the company. He was now filling various manufacturing and sales assignments for Ball Brothers. Burnham B. Holmes joined the company in 1950 as assistant to the president and soon would manage box and paper operations. In 1951, Kyle L. Menuez became general manager for rubber.

What the new personnel picture meant was that the operating committee, without a single member of the Ball family on it, was going to have a free hand in the day-to-day operations of the company.

Not that all central controls had been abandoned by any means. On questions of broad policy, the new team was to get its guidance from the Executive Committee: Chairman George A.

Ball, President Edmund F. Ball, with vice presidents John W. Fisher, Alexander M. Bracken, and George E. Myers, Ball Brothers vice president and treasurer since 1948.

The Ball Brothers' team had new blood and its best experienced leadership on hand for guidance and counsel.

There were vast and, in some cases, shocking changes in operational policies, and it took a great deal of cooperation and understanding on the part of both new and existing personnel to weather some of the storms and strife which would naturally arise when new manufacturing strategy and sales methods were explored.

But the in-depth management needed for the Ball Brothers' five-year program was on hand, in Ed Ball's words—"men with perspective, vigor, enthusiasm, and know-how."

Methods and processes were improved. New controls were applied. Market research gave the company a new perspective on sales. Field technical services to customers more than doubled. The whole system of warehousing and shipping was overhauled.

"We leveled off production peaks, and filled in the valleys, stabilizing employment," Ed said. "Properties which were not operating at a profit were sold or closed. Products that were lagging in sales were discontinued. Our wage and salary scales were increasing steadily and employees received a whole new program of benefits."

One of the major accomplishments of the rebuilding program under Menzies was the acquisition in 1952 of Kent Plastics Corporation of Evansville, Indiana, evidence that Ball was expanding its operations outside the company's traditional realms of business.

Menzies had heard about the plastics company

from Charles Inloe, president of National Bank of Evansville, who explained that Kent was operated by some able young men whose business had outgrown its financial capabilities. Upon invitation, Menzies took a look at it, was impressed and informed Ed Ball of the opportunity. Kent's customers included such blue chip interests as General Motors, Ford and Pepsi Cola and it was Ed's opinion the new company could do much to enhance the image of Ball Brothers. Menzies was given the go-ahead.

The company again became newsworthy because of its changing profile. *The Indianapolis News* reported on February 9, 1952, that while the firm and fruit jars had been synonymous for generations, "Today that is no longer true. Less than 10 percent of the sales of the far-flung corporation, which operates 12 plants in seven different locations, consists of fruit jars." The article listed the diverse products Ball Brothers made: glass containers for foods and beverages, corrugated paper and boxes, zinc battery cans and fruit jar caps, lithography for the printing industry and rubber parts for the appliance and automobile industries.

Ed Ball has recalled about the changing company image, "In spite of increasing awareness by several public institutions that the company was broadening its operations, there were many who continued to regard us as a pretty backward operation. There was even a lingering opinion that the company was a real dodo, still operating in the

Dark Ages. Word had not yet gotten around."

But word was spreading.

In September, 1953, an issue of *The Glass Industry* reported that the Ball Brothers company was competing in 17 container classifications and would likely broaden its field. Its great size and potential for continued growth prompted the magazine description of the company as a "sleeping giant." But that figure of speech really applied only to glass manufacture. When viewing the entire business, including zinc and rubber production, and the recently acquired plastics operations, the company could hardly be considered to be slumbering.

When Menzies left Ball Brothers in 1954, his departure came before the end of the projected five-year period of renewal that Ed had forecast. But the rebuilding process at least was begun under his management and the company, when he departed, was left in capable hands.

Burnham Holmes, who was active on the "team" to bolster Ball Brothers, today refers to Menzies' contribution as "an unfinished symphony."

"He never really accomplished all the things that he set out to do," Holmes comments. "But by the time he was gone only one to two years, it could be said that Ball was essentially a well-managed, modern company with the people generally knowing their jobs and doing them well."

Ultimately, Ball Brothers had replaced crisis with renewal.

Chapter 13

"THE NECESSITIES FOR CHANGE"

When Bill Schade came to Ball Brothers as executive vice president and general manager, to finish Duncan Menzies' "symphony," his low-key style was in stark contrast to Menzies' more flamboyant conducting. But his background was promising. Schade came to Muncie after several years as manager of an important division of Olin Industries of Alton, Illinois, and this was backed up with degrees in law and accounting.

At Ball Brothers, Schade accepted his responsibilities without fanfare and went to work in a businesslike manner, pulling together loose ends of the organization, establishing new procedures and improving systems of accounting. In this process Schade leaned heavily upon the assistance and advice of the management consultant firm of Booz, Allen & Hamilton, and upon a bright, young assistant, Howard Ward, in whom he had utmost confidence.

Several immediate shifts were made in the organization. Divisions were set up to operate more or less autonomously. John Fisher was appointed vice president for sales and Burnham Holmes was named vice president of the Allied Division under which came responsibilities for plastics, rubber, metals and closures. Ralph C. Edgar was vice president of personnel and public relations. Jack Bellinger became general manager of the Commercial Glass Division. Everett Ester was general manager of the Southwestern Division; Jim Donaldson continued as general manager of the West Coast division; G. Fred Rieman controlled government and industry relations. Schade and Ed Ball enlisted the counsel of experienced outside management executives to broaden the horizons of Ball Brothers' growth. John B. Place, then vice president of the Chase Manhattan Bank of New York, joined the board of directors in 1954 and in that capacity helped to improve the corporate image and provide an indication that it was no longer exclusively family dominated. He was also a strong supporter of corporate forward planning.

Changes were also made in corporate policy. This period of growth was to be concentrated on existing product lines, namely glass containers, rubber, zinc, plastics, and the need for establishing the nucleus of a research and development program was being studied. The "Toledo Decree" made it difficult to expand in the glass container industry, but available resources could be used to strengthen the drive toward diversification.

One of the traditional Ball product lines sorely needed attention. Substantial financial losses in the Muncie rubber department loomed as a major problem. The answer came when expansion in that field was completed that same year with the acquisition of the Chardon Rubber Company of Chardon, Ohio, in 1954. The plant had come to the attention of the firm through Kyle Menuez. He then worked with a Chicago business consultant or "finder," Stewart Cochran, to complete the details of the acquisition. Cochran played a major role in the company's reshaping by also directing Ball Brothers toward other acquisitions that would prove invaluable to its future growth. By coincidence, the new rubber company's major stockholders were members of the Bostwick family, three brothers of which were known to John Collett. After protracted negotiations, the Ball company purchased 78.5 percent of the common stock of the company for $1.3 million. Shortly afterward, responsibility for the Muncie operation was placed under the Chardon management with the combined operations reporting to Menuez. In time it was apparent that the Muncie shop could not be operated profitably. It was shut down and the machinery moved to Chardon.

The year following the Chardon acquisition, the death of the last of the original five Ball Brothers brought a great era to a close. George A. Ball was 92 when he died, his agile mind still intact. His life had spanned an important age in the development of his country and his beloved Midwest. His energetic and inspiring leadership would be sorely missed. The death of the chairman of the board was an augury of the company's coming emergence into the next generation of high technology and modern-day management.

Former President Herbert Hoover said at the time of G.A.'s death, "Mr. Ball's passing is a loss to his community and to the country. He was a leader of distinction and a man of great benevolence." Vice President Richard M. Nixon said, "In the death of George A. Ball, the country has lost one of its leading industrialists. Mr. Ball, together with his brothers, founded a company upon which a city largely depended; and, in the tradition of men of great wealth in America, he was lavish in gifts to philanthrophy and to foundations for religious and educational purposes."

Significantly, the same year that G.A. died came the move that resulted in the corporate leap, as Ed later described it, "from fruit jars to satelites." It was nothing G.A. could have imagined in the latter years of the nineteenth century when he made his arrival in Muncie in a railroad boxcar carrying his favorite horse. Yet significantly, it would prove to be probably the single most important accomplishment of Schade's years as general manager, and a rite of passage into fresh viewpoints.

Ball Brothers began, after World War II, to consider organized research essential to future growth. However, it wasn't until 1955 that the decision was finally made to establish a research and development department. A management consulting firm was engaged to analyze the needs of the company and found:

"At the present time Ball Brothers' research and development activity is largely limited to work in the operating units in answer to customer problems and a certain amount of process and product work farmed out to research and development laboratories. Almost none of this activity has resulted in major new products or process improvements. Furthermore, the new product lines which have been added are for the most part the result of acquisition of operating plants, and have not brought with them a formalized research and development force."

However, while the report criticized the company's past involvement in research, it stressed the need for further strides in the area:

"This study strongly indicates that a top level research and development function is needed at Ball Brothers Company to provide for:

1. Development of products with high profitability,

2. Bringing present process up to date and keeping abreast of industry,

3. Active planning and follow-up of research and development programs,

4. Greater engineering know-how, and

5. Improved morale among technical and supervisory employees."

As a result of this study, a careful search led to the naming of R. Arthur Gaiser as director of research and development. This man of unassuming manner and blunt speech would turn out to be an important asset to the company. His education included bachelor's and masters' degrees from Alfred University and advanced work at Buffalo University and Toledo University. His background was in ceramic engineering.

Originally aspiring to become a doctor, he steeped himself in the disciplines of medicine—biological chemistry, embryology, histology and anatomy. He also taught in New York state schools the mysteries of algebra, geometry, physics, chemistry and trigonometry by—as he says today—"keeping one chapter in the textbooks ahead of the students." That little masterpiece of understatement is typical, by the way, of a man who presently holds more than 40 U.S. patents, several foreign patents, and was the individual responsible for the development of glass capable of conducting electricity.

In brief, although at the time no one could foresee that before long the Ball company would be involved in the building of spacecraft or the creating of irrigating systems to make the desert bloom in North Africa, Gaiser just happened to be the right man to bring aboard to explore these new fronts of endeavor. It was to him that Ed gave the challenging assignment of developing a product for Ball (as Ed put it facetiously) that "could be made for a dime, sold for a dollar and was habit forming."

Gaiser was keenly aware of the firm's need for further expansion in research. Together with Bill Schade and Ed Ball, he carried on numerous conversations on the need to get into a more technically oriented business, not only for the sake of the business itself but also to show a skeptical public that the firm was forward-looking. His first job was to build and to staff a laboratory. This problem seemed a difficult task, to bring young dynamic researchers into an old line company, but the opportunities were so numerous and glaring that soon a working group had joined the firm and was producing results under the guidance of Dr. Christopher O'Shea. However, more sophisticated scientific disciplines were needed and here the company had another stroke of luck.

Ed recalls: "After looking at several firms which either didn't seem to fit or were outside of our reach financially, Cochran called our attention to a small electronics company located in Boulder, Colorado, known as Control Cells. He didn't particularly recommend it, knew nothing of the quality, practicality, or potential of its product, but it was small, appeared to have alert young men managing it, and we could look at it with no obligations. I later learned that this same company had been reviewed by Menzies a year or so before and he had turned it down, although quite rightly at that time."

It was Gaiser's task to make another evaluation

study of Control Cells during a "trial marriage" period of ownership. Gaiser reported that the company's principal product, a vehicle-weighing device, while a laboratory success, was impractical. Looking back, Gaiser dryly avers that anyone loading a truck would find this gadget very useful "if he also had a degree in engineering and was an expert in advanced calculus." Gaiser also recalls an ill-fated attempt by Control Cells to weigh a DC-6 aircraft right out on the airstrip, without taking into any consideration the effects of prevailing wind currents on wings and fuselage during the solemn weighing-in ceremonies.

Ball Brothers then turned to the nearby laboratories of the University of Colorado in a last-ditch attempt to evaluate the feasibility of the newly acquired device and reach a "go, no-go" decision on it. The answer, unhappily, was strongly negative, but, happily, the dead issue led to something more. Dr. David S. Stacy, who had worked with Gaiser on the device at the university, was then involved in other important research. While the U.S. space program was still a few years away, V-2 rockets had been brought over from Germany and Dr. Werner Von Braun was already on the scene applying his expertise in rocketry to the dream of placing men on the moon. The university had been building biaxial pointing controls for upper-atmospheric studies and Stacy had been involved in this program. The university at that time made a decision to concentrate on its prime function of basic research and get out of the "hardware business," thus freeing up some scientists and engineers. Ball Brothers seized upon a great opportunity. The company purchased Control Cells and employed these men in the manufacture of pointing controls and launch services. It also offered these men the challenge of applying their knowl-

edge to lead the company toward the frontiers of space technology.

Ball Brothers Research Corporation was formed under the laws of the state of Colorado in December, 1956. In its early months, "BBRC" was awarded additional NASA V-2 pointing-control contracts. Later, contracts were secured for a control to align a missile optically with the Earth's North Pole, balloon-borne pointing controls and infrared equipment, together with an instrument to measure ultraviolet light. Ultimately, these successful devices would lead to the contract award in 1959 for the NASA Orbiting Solar Observatory (OSO) satellite program, which was a crucial step in America's involvement in the "Space Race" of that era.

Writing a retrospective in *Nation's Business* magazine in 1973, then-company president John Fisher found the step "From the Kitchen to Outer Space" less surprising than it might have seemed to some outsiders. He wrote:

"Our experience in the space program after acquiring the technical company was in the tradition of the founders, who always sought to exploit fully any field they entered."

Meanwhile, the company expanded in other directions. In January, 1960, the company acquired Rolled Plate Metal Company of Brooklyn, New York, a major producer of zinc photo-engraving plates. Following this acquisition, a joint venture was entered into with the Grillo company of Frankfurt, West Germany, which also manufactured these plates. Eventually, the joint venture was terminated, but the experience was valuable because it familiarized the company with a system installed by Grillo known as the Hazelett Process for the continuous casting of zinc in large quantities, a system that would be utilized by the firm when it chose later on to consolidate its zinc operations.

In 1961 Ball Brothers purchased, for approximately $1.8 million, all properties and assets of the Industrial Rubber Company of St. Joseph, Michigan. The rationale behind its acquisition was that the plant would fit in with the Chardon rubber operation and give the company a substantial position in the mechanical rubber goods industry.

Other expansions accomplished by the Schade regime were in the traditional glass lines. Two entirely new plants were built in the years 1960-1961, part of a plan to have factories located geographically throughout the U.S. to offset increasing freight costs.

A southern site was chosen at Asheville, North Carolina, and a new glass plant opened in May, 1960. Asheville was chosen as a prime location because of its being in the approximate center of one of the most active home-canning supplies sales territories. Furthermore, one of the company's most valued commercial-ware customers had built a new plant in Asheville and other food processing plants were well within convenient servicing distance.

The second glass plant was built at Mundelein, Illinois, and opened in July, 1961, after an extensive search utilizing professional counsel for a site in the Chicago vicinity. While the recommended location satisfied most of the criteria specified, a serious oversight developed when it became difficult to develop a stable work force that would adjust to the rotating shifts of a 24-hour-a-day, seven-day-a-week operation, in spite of a study that showed an abundance of available labor.

Both projects were aided by much-needed funds which came from the sale of the Hillsboro glass facility in 1961 to a subsidiary of Hiram Walker & Sons, Inc. Ball had been a major supplier of liquor bottles to Walker and when the distiller planned to become a self-manufacturer, Ball seized the opportunity to avoid losses and, at the same time, to provide capital to finance its new glass projects.

The other Ball Brothers glass plants were doing well. El Monte, for several years, had been a profitable producer of high-quality glass containers. The Okmulgee plant, a small floundering operation purchased in 1929 from the Pine Glass Company, had inherited the best equipment from the obsolete Schram Glass Company's plant at Sapulpa, Ball's plant at Wichita Falls which closed in 1952, and the briefly owned and operated plant at Three Rivers, Texas. By the early 1960s it was a thoroughly modern plant and an important supplier of quality glass containers in the Southwest.

Consideration was given to an overseas glass container operation. Locations and possible joint-venture partners in France, Germany, and Spain were investigated. Even a firm in Casablanca, Morocco, which Ed had learned of during the war and later visited, was contacted. The building of a glass plant in Lagos, Nigeria, was also weighed, but several Ball executives visited Lagos and this possibility was dropped.

By January, 1962, President Edmund F. Ball was thus enabled to appraise the future of the company as favorable. He addressed employees in the *Ball Line*:

"The acquisition of Industrial Rubber gives us, with Chardon, a substantial position in the molded and extruded rubber and synthetic parts business and makes us an important supplier to the automotive, home appliance, and construction industries.

"With the completion of Mundelein, we believe we have in it and Asheville the two most modern and efficient plants in the glass container industry. El Monte, while older, has kept pace with technical improvements; and Okmulgee will soon undergo major repairs to furnace and equipment, thus providing us with four well-located, modern and efficient plants.

"Other divisions manufacturing zinc, metal and plastic products have operated well and we expect them to continue their growth....

"Prospects at Boulder Research are really exciting. Undoubtedly, the year will see one or maybe two satellites in orbit that were completely conceived, designed, and built in our Boulder laboratory....

"The year ahead will be a significant one. It will be filled with changes as must be all years ahead. There will be disappointments, there will be unexpected developments, and there will be opportunities. It will not always be pleasant or just as we would like to have things be, but there will be compensations.

"As I see it, Ball Brothers will continue to grow, become stronger, a better investment for those who have invested in it, and a better organization for which to work, just as long as it remains alert to the necessities for change and the accompanying opportunities."

An element of unease can be detected in Ed's column. References to "necessities for change," "disappointments," and "unexpected developments," however, may not have been readily understood at the time, but Ed was preparing them for what he would later call one of his most difficult decisions. "Properties not operating at a profit must be sold or closed," he had stated as his philosophy. The unproductive must not deny life to the productive.

The closing of the Muncie glass plant came soon after the operation had entered into its milestone 75th year. The huge old facility had a history of profit deterioration for almost 20 years and by the late 1950s it had become obvious to Ball Brothers management that some drastic action had to be taken. The problems were complex. The company had spent millions of dollars during the 1950s to modernize the plant, specifically in reconstructing the buildings and furnaces and replacing obsolete equipment, but problems were not solved. Its property taxes were high, labor costs were excessive and material handling was expensive. Losses climbed so high that by 1961 it was abundantly clear that the plant, once one of the world's largest glass container facilities, could no longer be operated profitably.

On December 13, 1961, the company's board of directors held a special meeting and plans were made to phase out production in the near future. The plant was to be closed entirely on March 15, 1962. Advance knowledge of these actions was confined to board members and a few executives. Then, on Saturday morning, January 13, 1962, the company invited about 50 community leaders to the corporate offices.

Ed Ball announced the decision, saying: "After long and careful consideration, Ball Brothers Company has reluctantly decided it is necessary to shut down its glass container manufacturing operation at Muncie.... The paper box division, which has functioned as a service department to the glass division, will also be closed down.

"This action will discontinue the jobs of several hundred employees....

"Our zinc-rolling mill and metal products division will continue to operate. Mold making and machine repair work will continue in part. The research laboratories will continue. Most of our administrative, engineering and clerical employees will be retained, as Muncie will continue to be the headquarters of our company. Thus the company will continue to give employment to some 800 men and women at the Muncie location.

"Discontinuance of the glass container manufacturing operation is necessary for economic reasons. Since the close of World War II this operation has been marginal or lost money. Every possible means has been used to put it on a sound, profitable basis. We have spent many millions of dollars for modern equipment and have introduced the latest methods of manufacturing.

"But there are a number of adverse factors that we have not been able to overcome. Included are an unfavorable plant layout that necessitates excessive material handling, old buildings that are expensive to maintain, worker inefficiencies, quality problems, employee costs that are higher than in competing plants, and high shipping costs to the major markets for the products we make.

"The glass container industry is highly competitive. In recent years, several older plants have had to drop out of the race. Regretfully, the Muncie plant must join them. . . ."

Ed added that he realized the impact the shutdown would have on displaced employees and that "we are taking action to do what we can to help them in the transition period while they are obtaining other employment." He said that the closing of the operation would serve to "strengthen the company and make more readily possible the realization of our plans for substantial expansion in the future."

The closing of the plant went surprisingly smoothly. Some laid-off employees went to work elsewhere in the Ball company while others went to work for the Corrugated Container Company which was located through Ball Brothers' efforts in warehouses remodeled for the purpose, utilizing equipment idled by the Muncie Box Plant. Voluntary termination benefits were negotiated with the appropriate unions and early retirement was worked out for those who were eligible. Ball Brothers called in from retirement a former personnel executive to establish a special employment office to help released employees who were having problems finding jobs.

In late 1963 an arrangement was negotiated with the Muncie Warehouse Company which agreed to occupy the enormous warehouses. Buildings considered undesirable for future use were razed. The six glass furnace stacks were also razed. Equipment was either transferred to Ball Brothers' other glass plants or, in most cases, sold to other companies.

However, the general emotional response to the action was not so controlled. Most Muncie residents were shocked to learn that what seemed for 75 years to be a permanent fixture in Muncie was closing down. Some residents were hysterical. Some of the labor unions and their members were publicly critical and voiced their dissent in full-page advertisements blazoning forth in the Muncie newspapers. The president of the international office of the Glass Bottle Blowers Association Union issued statements denouncing the company's action and filed a lawsuit that was dismissed without trial.

The frenzy, panic and shock seemed to center around either a misunderstanding of the announcement or a failure to deal realistically with the news. Many people mistakenly thought that Ball Brothers was forsaking Muncie by closing down all its operations in the city. Others, who at least appreciated the limited scope of the announcement, saw a more symbolic significance in the fact that Ball fruit jars would never again be made in Muncie. The news to them seemed to spell doom for fruit jars, Ball Brothers and Muncie.

The company did its best to re-explain the situation to confused and irate residents, but the controversy lingered. More than a year after the plant shut down, the situation continued. On October 13, 1963, a now-defunct newspaper, *The Indianapolis Times*, carried a story which was considered irresponsible and even damaging to the company and the community. Vern C. Schranz, director of public relations at Ball Brothers, chose to write a letter to the editor of that paper to apprise him of some facts, and, hopefully, at last to put the issue to rest:

"Mr. Editor:

"Your paper, last Sunday, carried a story regarding Muncie which left the impression that Ball Brothers Company had shut down all of its facilities in this community.

"This is not correct. The reporter who wrote the story either did a poor job of gathering his facts or he is quite inept in communicating the information he did secure.

"Here are the facts:

"1. Ball Brothers Company still employs nearly 800 people in Muncie.

"2. The Ball general offices in Muncie are the headquarters of a diversified and nationwide complex of industries. From this headquarters are directed the activities of the firm's 11 divisions, two subsidiaries, and one affiliate which operate manufacturing facilities in 15 communities in nine states, plus one in West Germany which will

soon begin operation.

"3. Ball production facilities in Muncie include (a) a zinc-rolling mill which is reputed to be the largest in the world, (b) a zinc drawing and stamping works, (c) a metal closure plant, and (d) a sealing compound manufacturing installation.

"4. Ball also maintains in Muncie one of the more extensive research and development facilities in the state. In these modern laboratories, skilled scientists and technicians are helping to lay the foundation for the company's future as they search for and find ways to improve current products and develop new product possibilities.

"No, Ball Brothers Company has not closed down its Muncie operations ... and with a single phone call your reporter could easily have determined the facts. ...

"V.C. Schranz"

The letter could not entirely put the issue to rest, but it did serve well to show the attitudes prevalent. Time, of course, would take final care of the glass plant crisis, but it seemed a mighty long time to those who had to bear the brunt of the criticism.

When Mr. Schranz' letter appeared in Indianapolis, another "letter to the editor" appeared in *The Muncie Star* at about the same time, written by the wife of a long-time foreman in the Shipping Department. It reflected the sorrowful aspect of the plant's closing, doing so without the usual hysterics:

"A friend has gone. Its voice has been silenced forever. For years I listened for the Ball Brothers whistle to set the clocks, to get the children up for school, used it to gauge the time to get the meals, etc. It has been my friend since I came to Muncie as a young bride in 1923. All through the years of the Depression, when everything looked so dismal

and hopeless to so many people, 'Old Faithful,' Ball Bros. whistle, went right on blowing. It was the one thing that could be depended upon in those dark days. I will certainly miss the old whistle, as much as the passing of a close friend or relative. I wish I could have done something to keep my old friend from passing into oblivion.

"But all good things must pass away it seems. It has gone the way of Ball Brothers glass factory, which was a part of the Muncie scene for so many years. So I shall bid a fond farewell to my friend, Ball Brothers' whistle, and say, all's well done."

Mrs. John Langdon
1421 S. Shipley Street

Part Five

From Muncie to the Moon

In the postwar period, Ball Brothers was beset by problems from many sides. Its traditions were strong, but its methods needed updating. Furthermore, the traditional fruit jar no longer could remain the backbone of the company. New directions were sought and the "present," a tidal wave thrusting an old-fashioned company into an immediate future of demanding technology, called for precedent-shattering decisions. The company would survive, but survival would tax the best qualities in the new generation of leadership. Progress lay in an orderly and predictable series of acquisitions and expansion—in glass, rubber, and zinc production. Ball Brothers' launch into space came as a dramatic surprise to the general public. "From Fruit Jars to Satellites" was a journey which Ed Ball and John Fisher *et al.* found themselves explaining over and over again in reasoned tones to the incredulous who found something faintly unreasonable in linking Muncie and the moon. But it was really not so much a revolution in company character as it was an evolution.

During its years of financial crisis, the firm turned to Duncan Menzies (left), who literally and figuratively moved under the eye of Ed Ball (right, same photo). Menzies described his aim as "to provide stability in production and employment." A human dynamo, he was also a buzzsaw with memorably sharp edges. He and other company executives moved dramatically to turn the firm around during this darkest period in its history. Ed Ball commented about the company during its crisis: "I often thought it might have been easier to have started an entirely new company than to restore an old one. But it still was a going concern with major assets, fine reputation and a willing organization."

"An operational team had been whipped together to put Ball Brothers back in the competitive race," reported *BusinessWeek* in February 1952. An integral part of the management team were two Ball in-laws: John W. Fisher (right), then vice president of metal and zinc closures; and Alexander Bracken (above), then vice president and general counsel.

A number of men made valuable contributions toward the continued success of the company past the years of crisis. Ed Ball (far left) remained a steadying influence throughout these years. Wilbert C. Schade (second from left) replaced Menzies as general manager and rose to the position of company president in 1963. At center is Burnham Holmes, a longstanding member of Ball's management group; R. Arthur Gaiser (second from right) who joined Ball as director of research and development in 1955; and Robert H. Mohlman (far right) who came to Ball as director of corporate planning.

Members of the Ball team gather for an informal session at the High Hat Restaurant. At table from left to right: Fred Rieman, Fred Kirk, Carl Richart, John Fisher, Bob Biggs. Standing from left to right: Van Fleet, Bill Fields, Wylie McDonald, Octavian Plymale, Sr., Louis Isselhardt, Ira J. Bird, Bob Campbell, Russell Sherward, Ed Flora, Don Daly, Francis Davis, Roger Burkett, Harold R. Clayton, Ed Shore, and Wilbur Hunt.

Zinc manufacture, dating back to the time Ball first made lids for its Mason fruit jars, continued to be an integral aspect of its business when the firm aimed to become more diversified. Today this Ball-made zinc roof of an Indiana department store demonstrates the varied uses to which the company has applied the versatile metal.

Ball also manufactures zinc for photo-engraving plates and for battery shells used by many of the nation's leading brand-name manufacturers. Continued growth in this area led to a metal decorating and service division which utilized other metals such as steel and tin, offering a wide variety of services. At left, youngsters power their precious toys, gifts from Santa Claus, with dry cell batteries to make them work, perhaps the most innocent power play in industrial history.

Ball had its start in the rubber business back when it was manufacturing sealing rings for fruit jars. With the acquisition of the Chardon Rubber Company in 1954 and the Industrial Rubber Goods Company in 1961, Ball was comfortably situated in a field, which, like zinc, was another outgrowth of its original home-canning business. A further expansion came in this field, indirectly, when Ball purchased Kent Plastics in 1952 and thereby entered a neighboring field that had exciting opportunities. Altogether, rubber and plastic products were manufactured for decorative and functional uses, most in the appliance and automotive industries. The illustrations on this page show the varied applications of rubber and plastic. At left, plastic packaging which was vacuum-formed comes off the production line at Kent Plastics.
Below right, more than 40 miles of Presto cove base moulding produced by Chardon Rubber were installed in the New York-Hilton Hotel. Below left, is the Memphis Airport Terminal where more than 10,000 feet of curtain wall gaskets were produced by Industrial Rubber Goods and now hold in place the terminal's 320 large windows shown here lining the top level.

The company's early efforts at research held close to technology which would better Ball's standard products and processes. Ultimately, the way pointed to an interest in a weighing device which, it was conceived, might be useful in calculating loads of raw materials reaching Ball plants. The device was owned by a small electronics company in Boulder, Colorado. While evaluating this possibility, Ball was suddenly confronted by a great opportunity in Boulder. University of Colorado scientists wished to go forward with pure research in the aerospace program and leave the manufacture of space hardware to private industry. Ball purchased Control Cells, Inc., and soon the former fruit jar manufacturers were making most advanced space equipment under government contract. The first of these contracts came in 1959 for a satellite program which proved a crucial step in the U.S. space plan. The NASA Orbiting Solar Observatory's task was to study radiation from the sun which might be a threat to future manned space flights. Six further OSO satellite contracts followed, but in retrospect, the first one was a NASA milestone and was later called "perhaps the finest satellite ever launched." At left, an OSO satellite lifts off, as one Ball advertisement put it, "asking the sun the way to the moon."

This is Control Cells' portable weighing device (above) which Ball's director of research R. Arthur Gaiser wryly declared would be useful to anyone loading a truck if "he also had a degree in engineering and was an expert in advanced calculus." The photo shows the platform of the device, which houses a weight-sensing cell. The separate indicator could be read at a distance. As an intrument for weighing, it was a failure—but it became a symbol of Ball's thrust into space.

168

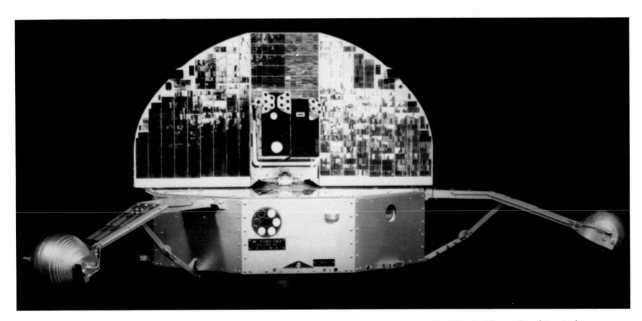

The OSO-1 satellite, above, represents Ball's first step in the manufacture of sophisticated aerospace equipment. Its task was to study solar radiation and how it might affect weather conditions on earth and manned space travel. A pathfinder for later space expeditions, it was called "a testimonial to the technical adequacy of the design and testing at BBRC." But the new challenge and its extensions seemed as illimitable as the skies. It began a program that made Ball a leading authority on solar observation and led to the development of a technological division which would encompass wholly new opportunities. As a prophetic BBRC report said in 1961: "It is highly predictable that from the aerospace industry will come new products which create new markets."

At right, Edmund F. Ball (left) and Arthur Gaiser (far right) converse with the distinguished space scientist (center) Dr. Wernher von Braun, director of NASA's Marshall Space Flight Center and a frequent visitor to BBRC. On this occasion, Dr. Braun was in Muncie to dedicate a new physical science, mathematics and nursing building on the campus of Ball State University.

While Ball became a more diversified company, it remained largely a container manufacturer. By the end of the 1960s, it had in operation four glass plants which were spread out across the nation, making commercial ware of all sizes and shapes and for such diverse products as shown at right. It broadened its base in the packaging field in 1969 when it purchased the Jeffco Manufacturing Company, makers of beer cans. Helped along by the development of a seamless aluminum can, the company grew rapidly under the guidelines of "strategic planning." The photograph below shows (from left to right) the step-by-step process involved in making these cans destined for picnic coolers and frost-free refrigerators across America.

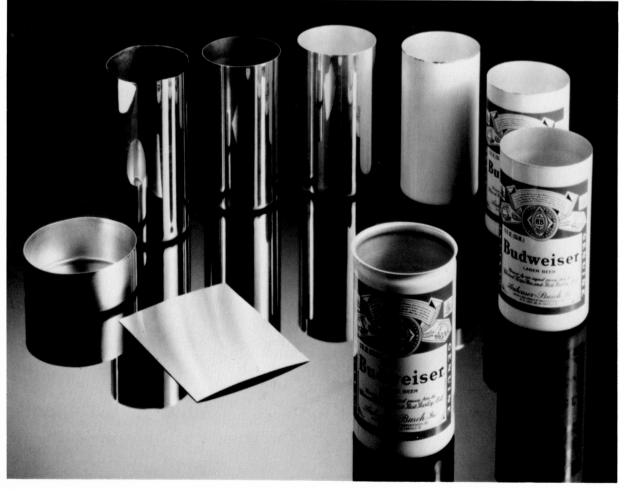

Chapter 14

"CHANGE IS A PROCESS, NOT AN EVENT"

Early on the morning of March 7, 1962, a hushed group of spectators stood on a small Florida beach and watched from a distance with awe something "gleaming white" (according to one newspaper's description). A rocket rose majestically from an inferno of flame and smoke and roared upward toward space and the fulfillment of its mission. All of the onlookers were stirred by the event, some even moved to tears. This was the group of NASA officials, Ball management, technicians, engineers and their wives and families who had worked day and night for two years to make this event possible. This was the culmination of their plans, efforts and skills. This was zero hour.

The event was the Cape Canaveral launching of the new OSO-1 satellite (Orbiting Solar Observatory), the highly sophisticated spacecraft Ball Brothers had conceived, designed and built for the National Aeronautics and Space Administration (NASA) under a grant of $1.4 million. Among the spectators was one surely more emotionally involved than many of the others: Ed Ball.

No doubt he felt wonder, and relief, too, as the craft soared into space. But Ed must have also felt that, in this moment, a prophecy was being borne out.

The prophecy that Ed was seeing partially fulfilled was one he made two years before in an address to the Newcomen Society, indicating that it could well be Ball technology that would help U.S. advances in aerospace, leading eventually to the placing of a man on the moon and the exploration of other planets. As a Ball advertisement of that time said: "OSO is asking the sun the way to the moon." The OSO-1 was now, in fact, a unit of the larger U.S. aerospace plan that preceded Neil Armstrong's first step on the moon. If man ever was to travel beyond Earth's orbit, more data would be needed on what to expect there. The Ball Brothers' satellite was crucial to that progress.

The satellite had been two years in the making and was especially significant because of its task to study potentially hazardous radiation given off by the sun, how this affects weather and other conditions on Earth and how much of a threat it could be to manned space travel. OSO-1 was designed to take the first direct look at the sun, undistorted by the earth's atmosphere.

In July, 1962, *BusinessWeek* heralded the event with the headline, "Fruit Jar Maker Scores in Space." The article read in part:

"From Cape Canaveral last March, the Nation-

al Aeronautics & Space Administration launched a sophisticated scientific satellite built by Ball Bros. Co., Inc., and the newspapers made sport of the space work of the 'fruit jar firm' . . . .

"It wasn't the kind of company that's generally associated with Cape Canaveral or the highly technical, multibillion-dollar U.S. space effort. Yet its subsidiary, Ball Bros. Research Corp. of Boulder, Colo., little more than five years old, has managed to climb to 20th place in last year's list of NASA contractors.

"Ball Research did it almost entirely because of its work on the satellite. If the little satellite, circling the Earth but focused on the center of the sun, can transmit for even part of [two months] it will go down as one of the most productive U.S. space research projects ever.

"OSO-1 was scarcely aloft when NASA gave Ball a contract for a second, even more complicated observatory, OSO-2. Ball also has a contract with the Navy to work on digital television, in which the picture is made up of dots, as in a newspaper halftone, rather than of lines, as in conventional home TV. Digital television is considered a better means of transmitting data from space satellites because its pictures can be fed directly into computers for analysis; it can also easily be scrambled and unscrambled as a means of transmitting in code. . . .

"Ball is drawing up a proposal for another major satellite contract, a larger and more ambitious solar observatory known by NASA as the S-67. Success in bidding for this contract could leapfrog Ball up among the leaders in the U.S. space effort."

Success indeed came amply. NASA lauded Ball Brothers for its work on the project and commended the research team for "a superb job." The satellite's performance, NASA announced, "is a testimonial to the technical adequacy of the design and testing at BBRC." Walter Orr Roberts, director of the High Altitude Observatory at the University of Colorado said of the project, "It is not secondary even to the manned space flight of Colonel John Glenn. This instrument will provide scientists with a spectacular advance of knowledge." Looking back, Space Administration officials proclaimed it years later "perhaps the finest satellite ever launched."

Ball Brothers' triumph with the OSO satellite led to the award of additional contracts for future OSO satellites, seven in all. Other space equipment BBRC was active in developing included solar sensors, pointing controls and processing lubricants to be used on metal spacecraft parts.

Then tragedy struck the project April 14, 1964, on the OSO-2 satellite. The third stage of the Delta rocket for launching the satellite was accidentally ignited by static electricity when joined with the satellite for a spin test. The satellite itself was severely damaged, and, worse, eleven engineers and technicians were burned in the explosion. Two Ball employees and a NASA technician died as a result of this accident. The Ball people were Sidney J. Dagle, engineer, and L. D. Gabel, machinist. It was a major blow to the space scientists working at Boulder and to all associated with the project. The only course to follow was to salvage all components and experiments aboard the satellite capable of being rebuilt and to attack the task of preparing another spacecraft for launch. In tribute to the men who lost their lives during these pioneer days in the Space Age, their names were put on a special plate that was attached to the subsequently launched satellite. Because of the highly delicate nature of these satellites, the memorial plate had to be positioned carefully so as not to upset the satellite's sensitive balance.

In March, 1963, Ed Ball came to a decision that seemed to cap the company's extensive period of renewal. He announced at the annual stockholder's meeting that he was relinquishing the company presidency while continuing on as chairman of the board. The man chosen to be the first person outside the Ball family to hold the position of president was Bill Schade. Schade would be the chief executive officer in charge of operations, and Ed would direct policy and formulate basic objectives at the board level.

Ed remembers: "I felt that this move was justified in view of the showing of the past nine years since Schade had come into the picture. Generally, I was satisfied with our progress and felt that Schade was entitled to the promotion. It seemed obvious from the assignment of duties that 'family' members were moving into more advisory positions and out of the line responsibility. That was the way I had planned it."

The move came as a surprise to company observers, not only because of the acceptance of a non-family member as the company's leader, but also because of the selflessness of Ed, who was fully capable of continuing in the job. But it was a further act of courage from the same man who assumed the job in the post-war period and who brought the company through the last and most painful chapter of the crisis, the closing of the Muncie glass plant.

Ed had decided, in his quiet way, that it was time for a change.

Approximately nine years after the transition from a period of crisis and Menzies' "strong medicine" treatment, it was becoming increasingly evident that the company was approaching another phase in its long and ever-changing history. It was no longer an out-of-date fruit jar business struggling for survival. It was becoming of interest to financial institutions as a company to watch, a substantial, diversified firm with, in addition, an exciting future in space exploration. Although it was still one of the few "name" companies privately owned, the truth was that through estate settlements, rapidly proliferating families, and stock options of employees, it was approaching the structure of a publicly owned company. It was becoming a more sophisticated company, too, with an increased awareness of its potentials.

The firm expanded notably during the first year of Schade's tenure as president. In January it became 50 percent owner of Kent-Reliance Limited of England. Plans were made to double Reliance facilities and equip it to produce the same quality of functional and decorative plastics made in Evansville, Indiana, for the automotive and appliance industries in Europe.

Another foreign adventure came in July, when President Schade reported to the board of directors on discussions with Glasindustrie Dongen NV of Dongen, Netherlands, regarding the possible establishment of a joint venture to build a new glass plant there. Its purpose would be to acquire a position in the packaging industry in the European Common Market. The venture was ultimately approved.

In April of 1964 it was revealed to the board that Caspers Tin Plate Company in Chicago, largest independent producer of lithographed metals in the United States, might be available for purchase.

It was not an unpredictable move for Ball Brothers to consider the acquisition. For many years, in fact back to the time of Frank C. Ball several years before World War II, when it had finally been determined that the old reliable zinc cap business was going to be taken over by the two-piece tin lids such as Kerr Manufacturing offered, Ball Brothers had a supplier-customer relationship, purchasing their tin-plate requirements from Caspers Tin Plate. Caspers also operated, adjacent to its lithographing plant in Chicago, the Lafayette Steel and Aluminum Corporation, serving customers in the Midwest with tin mill and sheet mill products and performing shearing and slitting operations. Both these businesses were obtained through the acquisition in August, 1964. The companies were combined and operated as a subsidiary until 100 percent of the stock was acquired in May, 1972, when it became a division of the company.

Ball Brothers Research Company was given a boost with the purchase of a Minnesota electronics firm, Miratel, in 1967. Miratel was a manufacturer of color and black-and-white television monitors for commercial, educational and cable television and would help in the expansion of BBRC's broadcast television equipment division. Miratel had a fine reputation, having pioneered the use of transistors in studio television monitors. Plans for future growth were in word-processing equipment, specifically in making monitors to be used in video display terminals. The division expanded rapidly after being acquired by Ball and eventually would become the company's Electronic Display Division to be located in new headquarters in Blaine, Minnesota.

Changes in personnel had a significant impact on the company in the 1960s. Several board directors served the company with distinction during these years, among them Alvin Owsley, Jr., a partner in a major law firm in Houston, Texas, who replaced his father as a director of the company; Reed Voran, a Muncie attorney and long-time legal counsel for the corporation; and Arthur M. Weimer, retired dean of the Indiana University School of Business and a prominent consulting economist, who was especially helpful in bringing specialized management techniques to the attention of the company. Weimer's sense of humor and storytelling abilities were always greatly enjoyed by his fellow board members. John B. Place, a vice president of Chase Manhattan Bank, soon to become its vice chairman and eventually to assume the presidency of Anaconda, brought, in addition to his financial expertise, a knowledge of and commitment to the value of corporate forward planning. Lester G. Porter, retired president of Borg-Warner, contributed his long experience as a corporate executive and was particularly helpful in analyzing operations and negotiating acquisitions. Robert Gwinn, chairman and chief executive officer of Sunbeam Corporation, was a valued consultant in many areas. William M. Ellinghaus, eventually to become president of AT&T, was also an important asset, constantly insisting, "If you do anything, do it first class." Robert H. Mohlman came to Ball Brothers in October, 1966, as vice president of corporate planning, having twenty years of experience in the container industry, specializing in finance and operations. His "keen perspective and analytical mind," as Ed called it, were considered important assets. Long-time aide Burnham B. Holmes, who had joined the firm as assistant to the president and then served in many different capacities, was designated a corporate vice president in 1968.

He also joined the board of directors, as did R. Arthur Gaiser, vice president for research and development. This combination of "outside" and "inside" directors brought an effective balance of judgment in decision making at the board level.

Such dramatic and significant advances as the thrust into space and the changing management influences signaled that the company character was changing, reflecting Ed Ball's philosophy that change is a process, not an event. Inevitably, such change created new problems, but now these were problems of success rather than problems of crisis.

The concept of a division of responsibility between representatives of stockholders (family) and the operating executives nonetheless created an overlapping of responsibilities and interests. Concepts of priorities differed. The pattern of growth from within established industries seemed clear; how to proceed became controversial. But it was healthy controversy. The company was no longer solely dependent upon one industry. It was now "a composite of related industries," as some observers called it.

Management faced decisions as how to best use financial and technical resources for growth instead of for mere survival. It could be called a period of selective opportunism.

There were nagging trouble spots in the latter half of the 1960s. What, for example, should be done with the Dongen glass plant, which was requiring a far greater commitment of corporate assets than had originally been anticipated? Worse, there were failures as well. A venture into a panel building construction program proved to be impractical and was abandoned. There was a grand plan to develop a chain of small plants to produce pre-sensitized engravers' plates located in strategic areas throughout the country that was scuttled by the unanticipated inroads of plastics. The joint venture in Germany with Grillo to produce zinc engraving plates for the European markets failed to find its anticipated markets and was eventually liquidated.

A merger with a company possessing ample resources to comfortably finance all of Ball's foreseeable projects was seriously but briefly considered. The same objections that F. C. and G. A. had recognized many years before when they considered a similar course were the determining factors. Who would manage the combined operations? Would Ball's identity be lost? What would become of faithful employees? The final decision was the same—the company would continue independently and be responsible for its own destiny.

While considerable gains were being made in sales, manufacturing processes and operating efficiencies, there seemed to be a lack of specific planning of the company's future. Ball had been growing through carefully considered opportunism during the 1960s. Now, approaching a decade of promise, the board and other members of the management team felt that the company's progress could best be assured by more formalized forward planning.

The investigation and establishment of strategically planned goals, objectives, and standards of performance were strongly recommended. Ed was charged by the board with the responsibility to investigate and recommend a course of action for the establishment of a corporate planning policy. After careful consideration, Ed's recommendation to the board was that McKenzie & Associates be engaged to develop a formal long-range strategic planning program.

Now only two years from the mandatory retirement age he himself had established, Ed resumed temporarily the position of company president. Schade had an excellent record of accomplishment as the first non-family president of the company, but he stepped aside in 1968 when differences of opinion developed over policies pertaining to corporate planning. Ed remained on as chairman of the board as well, and since his tenure as president was limited, he would be able to participate objectively in the planning and continued growth and progress.

Among other accomplishments, progress during the years of Ed Ball's second presidency came in the long-familiar zinc business. The company now owned three zinc-manufacturing facilities—in Muncie (where battery cans were made), Greencastle, Indiana (for lithographer's plates), and Brooklyn, New York (for photo-engraving plates), plus the joint venture in West Germany. Ball was the largest producer of rolled strip zinc in the United States and probably the world. It was discovered in 1959 that a process known as the Hunter continuous casting system, whereby molten metal could be drawn into a series of rollers, was operating successfully in California. It seemed possible that the same process might be used for casting zinc. Burnham Holmes negotiated a deal in 1960 and the new equipment was installed in the Greencastle plant. By 1962 it was operating successfully and produced a high-quality product. Greencastle thus became the first commercially successful continuous zinc-casting operation in the world. Another first for the company, it was an accomplishment in the zinc-rolling industry approaching in significance the earlier developments by Ball in the manufacture of glassmaking machines. With the Hazelett process, a higher volume procedure, another element for combined operation became feasible.

All things seemed right during the middle 1960s for combining the zinc operations into one highly efficient manufacturing unit utilizing this process. Sales volume was ample and sufficient tonnage was available to warrant installation of efficient, large volume equipment. Furthermore, the suggested site of the new facility, Greeneville, Tennessee, was considered excellent due to proximity to customers and the availability of labor, raw materials, power, and water. When the new facility was eventually built, starting in 1968 at a cost of $12 million, it was the largest single investment the company had ever made. Production began in 1970 and eventually the plant became one of the basic reasons for later "going public."

In 1969 came what Ed Ball would call "one of the most significant and successful acquisitions the company ever made," the purchase of Jeffco Manufacturing Company.

Caspers Tin Plate played an important role in Ball history at this point, having had a close relationship with a customer located in Nebraska City, Nebraska, known as Morton House Kitchens. It was operated by a group headed by Karl Nelson, who would later become a Ball director. Caspers had been decorating metals for Coors for its brewery located in Golden, Colorado. Several years before Ball's acquisition of Caspers, Coors invited the firm to set up a two-piece can manufacturing plant at or near Golden, but it was unable to do so because of its shaky financial condition. Caspers then referred Coors to Karl Nelson and his associates at Morton House, who were able to sponsor the project.

To start the operation, Nelson brought in Daniel Gabrielson, who resigned as a plant manager of Continental Can Company to help launch Jeffco in 1961. The company began with only 15 employees working within a small leased building, producing plain shells and ring-pull ends for the beverage industry. Soon three-piece, tin-plated steel cans were added to the product line, followed by bottle crowns and then by two-piece drawn and ironed aluminum cans. But success did not come suddenly. Until 1965, Jeffco grew with its only customer being its neighbor, the Coors brewery.

Ball Brothers acquired the company in 1969 as part of a merger through exchange of stock negotiated by Burnham Holmes with Morton House Kitchens. The merger included the acquisition of Alpine-Western, a companion firm organized to develop, engineer, and sell equipment used in can production, such as high-speed drying ovens for interior coating of cans and special washers for aluminum cans.

The same year of the merger, a twelve-ounce aluminum can line went into production. One year later, Jeffco's expertise in making the seamless, recyclable aluminum can was recognized by the Joseph Schlitz Brewing Company, which then contracted with Jeffco to engineer, design and assist in construction of an aluminum-can-making facility in Milwaukee, Wisconsin.

Prior to the introduction of formalized "strategic planning," no one except the executive committee, accountants, and auditors knew much about the company's finances. Art Gaiser remembers that at one of the early management meetings Ed Ball made the startling statement that in the previous year the company made just enough money to pay its electric bill.

But strategic planning brought new life to the firm. It was announced to company stockholders:

*"During the year a detailed 'Strategy for Growth' was prepared by management with assist-*ance of a leading consulting firm, and approved by the board of directors. The corporate objective is to establish this as a growth company whose financial performance will consistently equal that of America's largest and best-managed corporations.

"Specifically, the company has two major long-range objectives that represent minimum expected performance:
1. Perform in the top quarter of American business which will require the company to:
(a) Increase earnings per share by 10 percent or more each year.
(b) Maintain a return on shareholders' equity of 15 percent or more per year.
2. Develop a reputation in the financial and business community as an effectively managed, aggressively growing corporation."

The plan provided for annual development of a five-year strategic plan with corporate and division managers working together to identify and evaluate available alternatives, set priorities, control division activity in accordance with plan, measure performance in terms of results achieved, and reward managers on the basis of success in meeting objectives. Company officials informed stockholders:

"The goals established are challenging and demanding. The company will build on its current base of business, allocating resources according to the magnitude and attractiveness of the opportunity as balanced with the risk involved. The company also will seek to acquire other companies that complement or add significantly to current operations. Finally, the company will maintain a steady upward flow of competent, qualified managers through an aggressive program of management development, and will support these man-

agers with adequate organization, information, planning and control systems.

The various divisions of the corporation were now generally operating on a solid base. Strategic planning became a way of life for the company and all its operations. Experienced and capable management personnel were in place. It would be a busy, sometimes difficult, but on the whole very productive two years.

A symbolic gesture took place during the last part of Ed's tenure as president. In July, 1969, Ball Brothers Company, Inc., changed its name to Ball Corporation. Ed explained the move by saying that *Ball Brothers* was appropriate "when we were a small Indiana firm making fruit jars for home canning. Today, however, we participate in a wide segment of American industry with manufacturing facilities in ten states and in Europe. We believe the name of *Ball Corporation* more accurately conveys the scope of our expanding interests."

When it would come time for Ed to retire at the mandatory age, both he and the company were ready for John Fisher to take over as president and chief executive officer. Fisher had many years of experience with almost all the divisions and operations of the company. His election as president would come in 1970, moving Alexander M. Bracken to the position of chairman of the board. Ed's new position was chairman of the board's Executive Committee, a valuable post that would continue to allow him to observe the corporation as it moved dramatically toward its second century of progress.

Ball Corporation had indeed come into its own with its new identity and an entry into a new era.

In his address to the Newcomen Society, Ed Ball's prophecy had been far-reaching when he said: "Ours seems to be one of the few accepted laboratory-proven concepts for a soft landing on the moon. Very possibly, some of our work and some of our devices will be used to land the first human being safely on the moon or another planet."

On July 21, 1969, Neil Armstrong became the first man to set foot on the moon. What Ed Ball had seen begun on a beach in Florida with the launching of the first OSO satellite was now completed when the "Eagle" landed on lunar soil. While Ball friction-reducing coatings and a radiation-measuring device played a significant role in the Apollo 11 mission, it was really discoveries made possible through Ball-developed experiments and devices that largely contributed to this historic moment.

It was a fitting climax to a story which had began so humbly in Buffalo those many years ago. The time had come, swiftly in terms of the accomplishment it recorded, when Ed could say again with justifiable pride, as he did in his address to the Newcomen Society in 1960:

"Founded by five able and energetic brothers, we who now operate this fine company are proud of our heritage, proud of our company's accomplishments, and proud to have the opportunities that its future seems to hold. It is composed of a family of diversified yet related industries, each of which has substantial growth potentials, lending strength and stability to each other as well as to the whole. . . . We believe our products serve useful and essential purposes in our country's economy. We hope that through research we will develop new products and improve our old ones in such a way that we may keep pace with this fast-changing world and share in the tremendous future that unquestionably lies before us.

"It's a long interesting road from a kerosene can to a moon landing, and it's a great enterprise that has grown from the components of the common-place fruit jar. If the five brothers could view our operations now, I am sure they would be amazed. I hope they might be equally pleased."

Directors of Ball Corporation
as of January 1, 1980

John W. Fisher
Chairman of the Board and
Chief Executive Officer,
Ball Corporation.

R. Arthur Gaiser
Corporate Vice President
(retired), Ball Corporation.

Richard M. Gillett
Chairman of the Board and
Chief Executive Officer,
Old Kent Financial Corporation and
Old Kent Bank and Trust Company,
Grand Rapids, Michigan.

Burnham B. Holmes
Senior Vice President
(retired), Ball Corporation.

Betty M. McFadden
President,
Direct Marketing Division, Jewel
Companies, Inc., Barrington,
Illinois.

Robert H. Mohlman
Senior Vice President and
Chief Financial Officer,
Ball Corporation.

Alvin M. Owsley, Jr.
Partner, Baker & Botts,
Attorneys, Houston, Texas.

Richard M. Ringoen
President and
Chief Operating Officer,
Ball Corporation.

Robert M. Spire
Partner, Ellick, Spire & Jones,
Attorneys, Omaha, Nebraska.

Delbert C. Staley
President and Chief Executive Officer,
New York Telephone Company.

Reed D. Voran
Partner, DeFur, Voran,
Hanley, Radcliff & Reed,
Attorneys, Muncie, Indiana.

Arthur M. Weimer
Special Assistant
to the President, Indiana University,
Bloomington, Indiana.

INDEX